Analog Communications Systems

Analog Communications Systems

PRINCIPLES AND PRACTICES

K C Raveendranathan
Department of Electronics & Communication Engineering
Government Engineering College, Bartonhill
Thiruvananthapuram 695 035

Universities Press

Universities Press (India) Private Limited

Registered Office
3-6-747/1/A & 3-6-754/1, Himayatnagar, Hyderabad 500 029 (A.P), India
e-mail: *www.universitiespress.com*

Distributed by
Orient Blackswan Private Limited

Registered Office
3-6-752, Himayatnagar, Hyderabad 500 029 (A.P), India

Other Offices
Bangalore / Bhopal / Bhubaneshwar / Chennai /Ernakulam / Guwahati
Hyderabad / Jaipur / Kolkata /Lucknow / Mumbai / New Delhi / Patna

First published 2008
Reprint 2013

ISBN: 978 81 7371 620 1

Cover and book design

Set in Times New Roman 10/12 by
OSDATA, Hyderabad 500 029

Printed at
Graphica Printers and Binders
Hyderabad 500 013

Published by
Universities Press (India) Private Limited
3-6-747/1/A & 3-6-754/1, Himayatnagar,
Hyderabad 500 029 (A.P), India

Dedicated to the memory
of
my parents

Contents

List of Figures

List of Tables

Foreword

Over the past two decades, electronic communications has witnessed exponential growth. Dramatic changes are visible in communication systems, such as VLSI realisation of communication chips, mobile telephony, and broadband wireless. These developments are usually covered in undergraduate level courses on analog communications, digital communications, telecommunications, data communications and communication systems. Among these, analog communications forms the first level course and covers the important concepts used in communications.

This book provides a broad introduction to analog communication systems, covering topics ranging from simple modulation schemes to their performance analysis in the presence of noise. It provides the readers an excellent insight into the classical methods of mathematical analysis and design of analog communications systems. By including an introductory chapter on MATLAB and a number of MATLAB illustrative examples and plots, Professor Raveendranathan has succeeded in integrating MATLAB with communications studies. I hope that the student community would find this book useful since it covers the complete university undergraduate syllabus and the contents are up-to-date.

Dr M Hari Sankar
Adjunct Professor, Department of IT
Indian Institute of Management, Kozhikode

Preface

This book is the outcome of several years of my teaching *analog communications* in various engineering colleges affiliated to the three major universities in Kerala. It is intended to serve as a textbook on the subject for an undergraduate course in engineering/science. Although analog communications has long been a part of the undergraduate programme leading to a BTech. degree in electronics and communications engineering, the thrust on using MATLAB programming environment as a backbone for simulating the techniques is quite new. It is rightly said that courses such as signals and systems, communications theory, control systems, analog communications, digital communications and digital signal processing can greatly benefit from the capabilities of this high-level language. MATLAB is extremely user friendly and offers a fast learning curve. MATLAB has been used in this book to illustrate concepts, simulate signals, and so on.

Although the book was written with Kerala University's syllabus of analog communications in mind, it can serve equally the needs of other universities as well. Numerical examples are given as solved problems, whenever possible, throughout the book. Additionally, each chapter has exercises at the end in which the answers have been provided for numerical problems. MATLAB source codes are provided for various tasks such as generation of signals; these have been tested on MATLAB version 7.0 (Release 14) and made nearly bug-free. They should also work well with a lower version such as 6.1 (Release 12), and readers are encouraged to try them and reproduce the results for themselves.

The book begins with a brief introduction to the concepts of analog baseband. The limitations of baseband communications are discussed. Chapter 1 also provides a brief look at the history of communications engineering; the important innovations (not exhaustive) are presented in a tabular format. The pros and cons of digital communications are discussed. It is expected that the introduction to MATLAB programming, included as an appendix to the chapter, will serve the student as a stepping stone to programming in this environment. For advanced learning, the reader is advised to refer to any of the books listed in the bibliography.

The second chapter is entirely devoted to telephone systems. Starting with the plain old telephone systems (POTS), the principles of telephone systems, which includes diallers, telephone circuits, toll trunks, and telephone switching fabrics, are discussed. The chapter also includes a discussion on recent trends in technology, the concept of circuit switching and basic switching networks (included as an appendix). A number of numerical examples are also given.

Chapters 3 and 4 are devoted to analog linear modulation. Chapter 3 begins with the concept of modulation, its requirements and merits; the mathematical analysis of the same follow. The treatment is made simple and lucid without, however, sacrificing the necessary rigour. Various forms of amplitude modulation (AM) are discussed and a comparative study of the same presented. Using MATLAB script, the amplitude modulated waveforms for various depths of modulation have been generated and plotted. Several practical circuits have also been provided. Some of the recent developments in AM, viz. digital amplitude modulation and DRM are treated in detail.

Angle modulation is the topic of discussion in Chapter 5. The mathematical analysis of FM and PM waves, an introduction to Bessel functions and the interrelation between FM and PM are dealt with in this chapter. Simulation of FM and PM generation is provided for typical input waveforms such as the square wave and the triangular wave. The hardware involved in modulation and demodulation are discussed. Chapter 6 deals with probability and random processes. Probability and random processes form a necessary part of analog communications, and students are expected to have a basic knowledge of the probability concepts. In this chapter, the concepts of random variables and random processes, statistical independence, autocorrelation and cross-correlation functions and covariance are introduced. A number of probability density/mass functions are introduced with the help of their mathematical models and the corresponding two-dimensional plots are generated using MATLAB. The means and variances of popular distributions are tabulated; this will greatly aid students in solving problems. Selected, graded problems are included as exercises at the end.

The analysis of various modulation systems (discussed in Chapters 3, 4 and 5) in the presence of noise is the subject matter of Chapter 7. The performance of AM and FM systems is covered in detail.

The last chapter is devoted to a discussion on electrical noise in electronic systems. Various noise sources are identified and their power spectral densities provided. Narrowband noise is analysed vis--vis its characterisation in communication systems. The Rician probability distribution arising from the joint probability density function of the envelope and the phase of narrowband noise along with a sinusoidal wave is obtained.

A selection of question papers from the examinations conducted by Kerala and Kannur Universities is included at the end of the book as a reference for students as also a detailed bibliography that includes several internet references. The book is amply illustrated with diagrams to make the discussions clear and lucid. It is the author's hope that the book will serve as good textbook and a reference to the interested reader. Even though much care had been taken in typesetting the book in LaTeX , it is possible that some errors have crept in. Readers are requested to communicate such occurrences to the author, however trivial they may seem.

Acknowledgments

The author would like to place on record his immense gratitude to his students, colleagues and the reviewers of this book who have contributed in various ways to bringing this publication to this form. I would like to mention in particular the contribution of my old classmate and lifelong friend, Dr. G. Abhilash of National Institute of Technology, Kozhikkode, who went through the manuscript critically and made several valuable comments. A book of this kind needs several hours of writing and rewriting and a lot of isolation from one's family. The author was fortunate to have the support of his wife, who was very patient and understanding about his academic goals . For that, I thank my wife Sobha and daughters Anuja and Aabha, who left several evenings/late nights and the study entirely at my disposal.

K C Raveendranathan
ravi@gecbh.ac.in

Chapter 1

The Analog World

1.1 Introduction

This book intends to introduce the principles and practices of analog communications systems. It begin with a brief introduction to the analog world. The purpose of any communication system is to convey information or message from one point known as the source, to another point, the destination. The message is usually transmitted in electrical form and is called the *signal*. A signal can assume *infinitely many* values. At this point it may be noted that a signal is, in effect, the electrical representation (or counterpart) of a physical phenomenon like the temperature or the wind speed at a point. The signal can be a (mathematical) function of many variables; time, for instance can be one of them. When the independent variable is time, the signal is often called a *temporal signal*; on the other hand, if the independent variable is space, the signal is called *spacial signal*. Speech and video signals are examples of temporal signals, whereas a still picture is an example for spacial signal. An example of a one-dimensional signal is the speech signal, where the independent variable is time.

1.1.1 Signals, media, and systems

In brief, any communication system needs a signal (which is the information to be conveyed in electrical form or other), a medium to carry the signal and a system which includes all the paraphernalia at the source and the destination ends which *effect the communication process.* We need the medium to convey the message from the source to the destination. Examples of signals are the speech or the voice signals in a telephone system, the radio/TV broadcast signals, etc. The corresponding media are electrical cable pairs or optical fibres and radio waves transmitted through free space.[1] Considering the various types of signals and the media for conveying these signals, we can have the following four combinations:

1. Analog signal and analog medium (Type-1).
2. Analog signal and digital medium (Type-2).
3. Digital signal and analog medium (Type-3).
4. Digital signal and digital medium (Type-4).

One can site the plain old telephone system (POTS) or the analog telephone as an example of a Type-1 communication system while the present-day mobile phone system is an example of Type-4. Similarly, we can find examples of other types of communication systems. This book

[1] The term 'Ether' was used to describe free space. The term 'Ethernet' has its origins in the term ether.

mainly focuses on analog communication systems, though we will have cursory discussions on the other types as well whenever needed. We will now elaborate the term 'analog' in the next few lines.

The term analog means something that has a continuous 'trail'. Thus analog signals are continuous in the time domain. This is against the discrete nature of the digital or discrete signals which are discontinuous in the time domain. Examples of typical analog signals are speech signals and the ECG/EEG/EMG signals. Discrete signals do not have continuity in the time domain *as they are not continuous functions of time.* One can note that certain signals are discrete by default while others are analog. An example in case is the ambient temperature of a room, which, the reader would have correctly guessed, can be expressed as an analog electrical signal by using a suitable *temperature transducer.* However, we can *discretise* the signal thus obtained by restricting the time of observation to *whole* hours. MATLAB plots of sample analog and digital signals are plotted in Figure 1.1.[2] Analog signals can be converted to digital form by using analog-to-digital (A/D) converters. Digital signals on the other had, can be converted back to analog form by means of D/A converters.

A point to be emphasised at this level is that the real world, as we know it, is de-facto and by default, analog. This is due to the reason that, as we will learn later, most of the naturally occurring information sources are analog; the channel (or medium) for transmission of signals is analog and so are their information receivers or the sinks! As you see, *it was all analog in the beginning, the rest is man-made.*

The MATLAB code for obtaining the plots in Figure 1.1 is given below.

```
>> x=[-10*pi:.01:10*pi];
>> y=[0:18, 17:-1:0];
>> subplot(211),plot(x,sin(x)./x),xlabel('Time t'),...
   ylabel('sin(t)/t'),title('Figure 1.1(a): An Analog Signal');
>> subplot(212),stem(y),xlabel('Time t'),ylabel('y'),...
   title('Figure 1.1(b): A Digital signal');
>>
```

1.1.2 Signal in the frequency domain

It may be noted that the signal, which is normally represented in the time domain, can also be represented in the *frequency domain.* We make use of various transformations of signals for this purpose: the Laplace transform, the Fourier transform or the z-transform. It is interesting to note that any periodic, real time signal with a finite number of discontinuities[3] can be represented by the sum of amplitude-scaled sines and cosines; this is known as the Fourier series representation. The transforms mentioned earlier get the frequency content of a signal: they represent the signal in the frequency domain. Quite often, the frequency domain representation of a signal, using any of of the transforms mentioned above, is referred to as the *spectrum of the signal.* Obtaining the frequency spectrum is quite useful and informative in the study of signals.

The type of transform to be applied to the signal representation depends on the nature of the signal. For example, Laplace transform and Fourier transform are used for signal analysis

[2]There is a subtle difference between the terms *discrete* and *digital.* In the case of the latter, the dependent variable (or function) is further approximated to an integer value.

[3]This is known as the Dirichlet's condition.

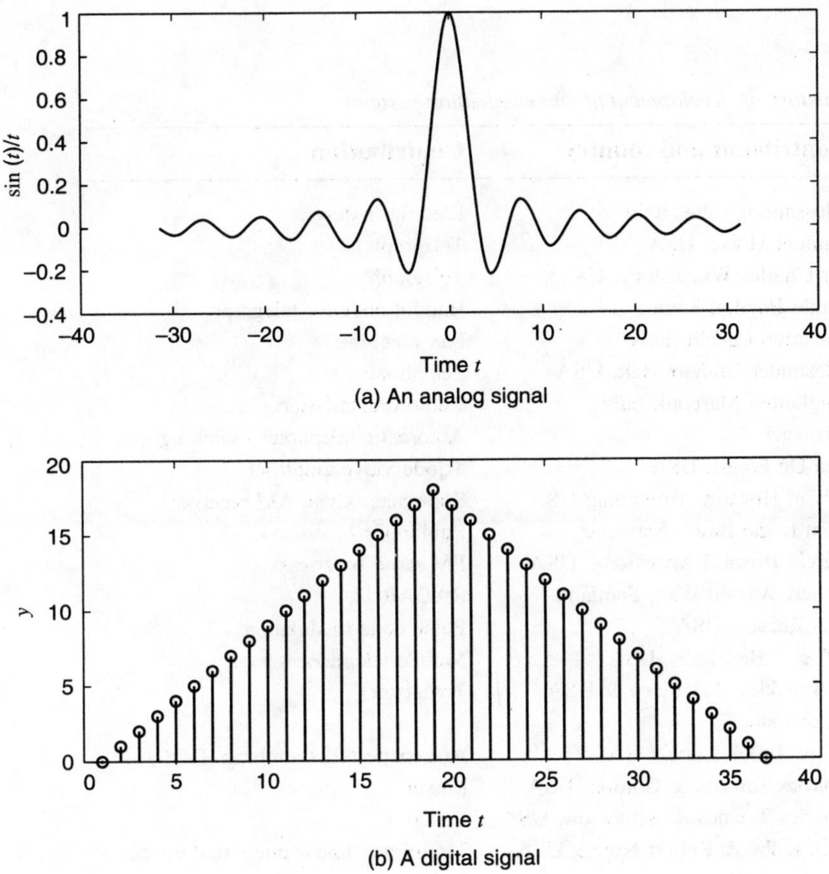

Figure 1.1 *Analog and digital signals*

of analog signals, whereas z-transform is used on discrete signals. Students can refer to any standard book on signals and systems for detailed information on signal transforms.

1.2 A Brief History of Communications

The earliest forms of communication were probably vocal-chord sounds generated by animals and human being and received by the ear. When the need for communcating over larger distances evolved, the sense of light was employed to augment the information content of the sound. In the second century BC, Greek telegraphers were known to have used torch signals to communicate. Different combinations and positions of the torches were used to represent various Greek alphabets. In the 18th century, communication of letters was accomplished using semaphore flags; however, their reliability was severely limited by the line-of-sight distances.

The communications scenario has undergone tremendous developments since then. Table 3.1 gives a chronological listing of technology development in the history of electrical/electronic communications engineering.

Table 1.1 *The history of development of communication systems*

Year	Contributor and country	Contribution
1799	Alessandro Volta, Italy	Electric battery
1835–1837	Samuel Morse, USA	Telegraph
1837	Sir Charles Wheatstone, UK	Telegraph
1860	Emile Baudot, France	Baudot code for telegraph
1861	Giovanni Caselli, Italy	Fax machine
1876	Alexander Graham Bell, USA	Telephone
1896	Guglielmo Marconi, Italy	Radio transmission
1897	Strowger	Automatic telephone switching
1906	Lee De Forest, USA	Triode valve amplifier
1920	Edwin Howard Armstrong, USA	Superheterodyne AM receiver
1926	John Logie Baird, Scotland	Public TV broadcast
1933	Edwin Howard Armstrong, USA	FM radio broadcast
1935	Robert Watson Watt, Scotland	RADAR
1937	H.A.Reeves, USA	Pulse code modulation
1946	AT & T Bell Labs, USA	Mobile telephone service
1947	W.Shockley, J.Bardeen & USA W.Brattain,	Transistor
1950	AT & T Bell Labs, USA	Automatic toll switching (DDD)
1954	Charles Townes & Gordon, USA	Maser
1958	Charles Townes & Schawlow, USA	Laser
1958	Jack Kilby & Robert Noyce, USA	Monolithic linear integrated circuit
1960	Sir Arthur C Clarke	Satellite communications
1965	C K Kao	Fibre optic communications
1969	Williard Boyle & George Smith, USA	Charge-coupled device (CCD)
1976	Robert Metcalfe & David Boggs, USA	Local area network
1980	Sony Corp., Japan	Experimental HDTV
1980	Philips & Sony Corp.	Compact disc ROM (CD ROM)
1984	ISO	Integrated services digital network
1990	Sir Timothy Berners–Lee, UK	The world-wide web (WWW)
1994	Sir Timothy Berners–Lee	World Wide Web consortium (W3C)
1996	Philips, Sony & Matsushita	Digital versatile disc (DVD)
1998	ITU	Facsimile over Internet protocol
1999	The IEEE Inc., USA	802.11a & 802.11b Standards
2002	The IEEE Inc., USA	802.15.1 Bluetooth PAN
2003	Intel, USA	Mobile Centrino CPU
2003	The DVD Forum	High density DVD (HD DVD)
2004	The Wi-Fi Alliance	Wi-Fi protected access

Graham Bell's First Call Named IEEE Boston Section Milestone

In a rooming house attic in Boston, in 1876, Alexander Graham Bell summoned his assistant, Thomas Watson, in the world's first telephone call. Now 130 years later, this first transmission of intelligible speech over electric wire has been designated an IEEE Milestone in electrical and computing engineering. The IEEE Boston Section celebrated this milestone on March 10, 2006.

Back in 1876, Bell and Watson felt they were close to a great discovery, and so they moved their laboratory from an electrical shop at 109 Court Street to a boarding house for privacy. Bell, a Scottish immigrant, taught deaf children and wanted to use electricity to help his students communicate. Some years before he invented the telephone, Bell, who had no formal technical training, started tinkering with the telegraph in the hope that it would transmit audible voices.

After Bell invented the telephone, he and his associates formed National Bell Telephone Co., predecessor of American Bell Telephone, in Boston, to lease handsets and offer telephone services to subscribers. For the next 30 years, Boston was the home of telephony research and development in the United States. In 1907, American Bell Telephone, at that time known as AT&T, moved its corporate headquarters and engineering operation to New York City.

(*http://www.theinstitute.ieee.org*)

1.3 Analog Communications System

The analog communication system in its simplest form consists of just three blocks: the message source, the communication channel (or medium) and the message sink(or receiver). A simplified block diagram of an analog or general communication system is given in Figure 1.2. We can learn about each block in detail as we go deeper into analog communication techniques. It may be noted that we transform the original electrical message by passing it through several stages in the transmitter before handing it over to the channel; these transformations are to be undone at the receiver.

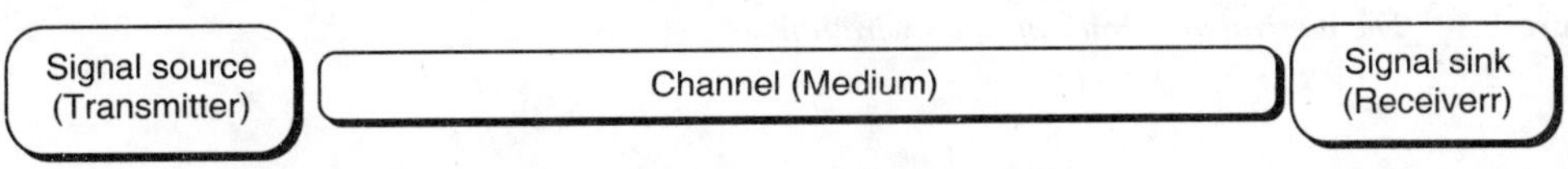

Figure 1.2 *Simplified block diagram of analog communications system*

The channel that links the transmitter to the receiver can be an electric wire, an RF channel or a fibre optic cable as in the case of a telephone system, or free space (RF or microwave channel) as in the case of radio or TV systems. It should also be noted that the source and sink need not be connected physically to have a communication take place! As communications system design matured over time, digital communications gained wider appeal due to several advantages it offered in comparison its analog counterpart. In digital communication systems, the analog input signals are first converted into digital form by *analog-to-digital converter* (ADC) and then transmitted over the channel. At the receiver, the received digital signals are converted back to the analog domain by means of *digital-to-analog converter* (DAC). We will now discuss some of the commonly employed channels of communication and then introduce digital communications system in some more detail.

1.3.1 Open wire line

A pair of open wires are used in the local loop of the plain old telephone services (POTS). The equivalent circuit of the open wire line consists of lumped parameters including resistors, inductors, and capacitors as illustrated in Figure 1.3. Typically, it can be shown to have lowpass filter (LPF) characteristics. The open wire line offers only a low bandwidth and is prone to various forms of electromagnetic interference as there is no shielding of conductors. The two-wire line is normally operated in the *balanced mode*, the conductors being arranged such that they present equal capacitances to the ground. The parameters R_s, L_s, C_{sh}, and R_{sh} are known as the *primary line constants*. No simple formulae that cover the entire range of frequencies can be given for the primary constants, but simplifications are possible for low or audio frequencies and high or radio frequencies.

For a two-wire line with the conductors embedded in a medium of permittivity ϵ (F/m) and permeability μ (H/m) with line dimensions given in metres, the primary inductance and capacitance per unit length is given by

$$L = \frac{\mu}{\pi} \ln\left(\frac{2D}{d}\right) \text{ H/m,} \tag{1.3.1}$$

$$C = \frac{\pi\epsilon}{\ln(2D/d)} \text{ F/m.} \tag{1.3.2}$$

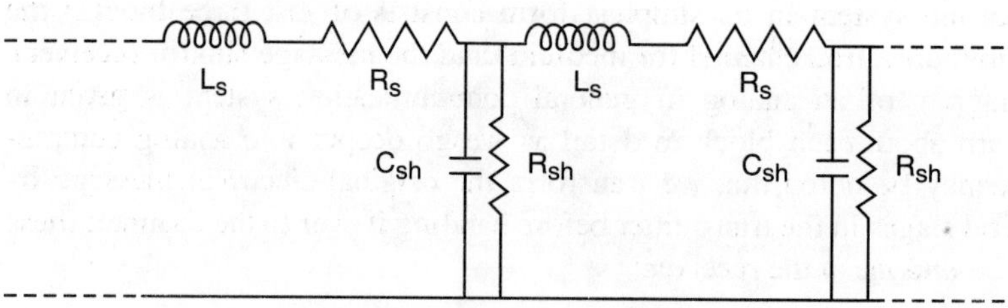

Figure 1.3 *The equivalent circuit of open wire lines*

1.3.2 Unshielded twisted pair cable (UTP)

The unshielded twisted pair is the most common kind of copper telephone wiring. Twisted pair is the ordinary copper wire that connects home and many business computers to the telephone company. To reduce crosstalk or electromagnetic induction between pairs of wires, the two insulated copper wires are twisted around each other. Each signal on a twisted pair requires both wires. Since some telephone sets or desktop locations require multiple connections, twisted pairs are sometimes installed in two or more pairs, all within a single cable. In some business locations, the twisted pair is enclosed in a shield that functions as ground. This is known as shielded twisted pair (STP). Even though UTP cables do not offer as high bandwidths or as good protection from interference as coaxial or fibre optic cables, they are extensively used in local-area networks (LANs) and telephone connections because they are comparitively less expensive and easier to work with.

Table 1.2 *Different categories of UTP cables*

Category	Cable type	Use	Bandwidth	Maximum data rate
1	UTP	Analog voice	–	–
2	UTP	Digital voice	–	1 Mbps
3	UTP, ScTP, STP	Ethernet	16 MHz	4 Mbps
4	UTP, ScTP, STP	Token ring	20 MHz	16 Mbps
5	UTP, ScTP, STP	Fast ethernet	100 MHz	100 Mbps
5e	UTP, ScTP, STP	Gigabit ethernet	100 MHz	1 Gbps
6	UTP, ScTP, STP		200 MHz	10 Gbps
7	STP, ScTP		600 MHz	

UTP: Unshielded twisted pair; STP: Shielded twisted pair; ScTP: Screened twisted pair

The different categories of UTP cables are given in Table 1.2. The standard connector for unshielded twisted pair cabling is the RJ-45 connector. This is a plastic connector that looks like a large telephone-style connector. A slot constrains the RJ-45 to be inserted only one way. RJ stands for Registered Jack, implying that the connector follows a standard borrowed from the telephone industry. This standard designates which wire will go with with which pin inside the connector. RJ connectors (different types like the RJ-11, RJ-22 and the RJ-45) are used at various points in a telephone circuit. For example, RJ-11 connectors are used to connect the telephone handset to the basic telephone equipment, while RJ-22 connectors are used to connect the telephone equipment to the wall-mounted domestic socket. In telephone applications, UTP is often grouped into sets of 25 pairs according to a standard 25-pair colour code originally developed by AT&T. A typical subset of these colours (white/blue, blue/white, white/orange, orange/white) shows up in most UTP cables.

1.3.3 Shielded twisted pair cable (STP)

A disadvantage of UTP is that it may be susceptible to radio and electrical frequency interference. Shielded twisted pairs (STP) are suitable for environments with electrical interference; however, the extra shielding can make the cables quite bulky. Shielded twisted pairs are often used in networks employing token ring topology. The shield around each of the wire pairs plus an overall shield protects the STP in noisy environments against excessive electromagnetic interference. A variation of the STP, known as ScTP (screened twisted pair) or FTP (foil twisted pair), uses only the overall shield and provides more protection than UTP but not as much as STP.

1.3.4 RF cable

The radio frequency (RF) coaxial cable, commonly called the coax, is widely used as a wideband communication channel. Coaxial cabling has a single copper conductor at its centre. A plastic layer provides insulation between the centre conductor and a braided metal shield. The metal shield helps in blocking outside interference from devices such as fluorescent lights, electrical motors, computers, etc.

Although coaxial cabling is difficult to install, it is highly resistant to signal interference. In addition, it can support greater cable lengths between network devices than can the twisted pair. Two types of coaxial cabling are available: thick coaxial (also referred to as thicket) and thin coaxial (thinnet). 10Base2 refers to the specifications for thin coaxial cable carrying Ethernet signals. The 2 refers to the approximate maximum segment length being 200 meters. In practice, the maximum segment length is 185 meters. Thin coaxial cable is popular in school networks, especially linear bus networks. 10Base5 refers to the specifications for thick coaxial cable carrying Ethernet signals. The 5 refers to the maximum segment length being 500 metres. The thick coaxial cable has an extra protective plastic cover that helps keep moisture away from the centre conductor. This makes thick coaxial a great choice when running longer lengths in a linear bus network. The disadvantage of thick coaxial is that it does not bend easily and is difficult to install.

Coaxial cable connectors

The most common type of connector used with coaxial cables is the Bayonet-Neill-Concelman (BNC) connector[4]. Developed in the late 1940s as a miniature version of the Type C connector, BNC is named after engineer Carl Concelman. The BNC product line is a miniature quick connect/disconnect RF connector. It features two bayonet lugs on the female connector; the mating is achieved with only a quarter turn of the coupling nut. BNCs are ideally suited for cable termination for miniature to sub-miniature coaxial cable (RG-58, 59, to RG-179, RG-316, etc.) RF cables with both 50Ω and 75Ω characteristic impedances can be used along with BNC connectors. Different types of adapters are available for BNC connectors, including T-connector, barrel connector, and terminator. Connectors on the cable are the weakest points in any network. To help avoid problems with in the network, it is advisable to use the BNC connectors that crimp, rather than screw onto the cable. The BNC connector is illustrated in Figure 1.4.

Figure 1.4 *The BNC connector*

Figure 1.5 *The TNC connector*

The TNC connectors are another type of connectors used along with RF cables. Developed in the late 1950's, TNC stands for 'threaded Neill Concelman'. A threaded version of the BNC, the TNC series features screw threads for mating. TNCs are miniature, threaded, weatherproof units with a constant 50Ω or 75Ω impedance and operate from 0 – 11 GHz. There are two types of TNC connectors: standard and reverse polarity. Reverse polarity is a

[4]BNC is wrongly expanded as 'Bayonet-Naval-Connector' in many citations, probably due to the 'bayonet' term and its military connotation.

keying system accomplished with a reverse interface; it ensures that reverse polarity interface connectors do not mate with standard interface connectors. Figure 1.5 shows the profile of a typical TNC connector.

1.3.5 Microwave link

Microwave links were commonly used in the long distance telephone trunk lines—links connecting distant central offices (telephone exchanges). But their use has been minimised with the advent of cellular radio channels and optical fibre links. Microwave links can of two types—terrestrial microwave links or satellite–earth-station microwave links. The repeater spacing in the case of the former is typically 60–90 km while in the latter it is of the order of 36,000 km.

1.3.6 Optical fibre cable (OFC)

Fibre optic cabling consists of a centre glass core surrounded by several layers of protective materials. It transmits light rather than electronic signals, eliminating the problem of electrical interference. This makes it ideal for environments that contain a large amount of electrical interference. It has been made the standard for connecting networks between buildings due to its immunity to the effects of moisture and lighting. Fibre optic cable has the ability to transmit signals over much longer distances than coaxial or twisted pair. It also has the capability to carry information at vastly greater speeds. This capacity broadens communication possibilities to include services such as video conferencing and interactive services. The cost of fibre optic cabling is comparable to copper cabling; however, it is more difficult to install and modify. 10BaseF refers to the specifications for fibre optic cable carrying Ethernet signals. Note the following with regard to optical fibre cables:

- Outer insulating jacket is made of Teflon or PVC
- Kevlar fibre helps to strengthen the cable and prevent breakage.
- A plastic coating is used to cushion the fibre centre
- Centre (core) is made of glass or plastic fibres

Fibre optic connectors

The most common connector used with fibre optic cable is the ST (simple termination) connector. The ST connector employs a plug and socket which is locked in place with a half-twist bayonet lock. The ST connector was the first, *de facto* standard for fibre optic cabling and has been standardised as FOCIS 2 (fibre optic connector inter-mateability standards) in EIA/TIA-604-02. It is barrel shaped, similar to the BNC connector. A newer, square-faced connector, the SC, easier to connect in a confined space, is becoming popular. It has been standardised as FOCIS 3 in EIA/TIA-604-03. Yet another connector, the LC connector developed by Lucent Technologies for use in TelCo environments, is a small form-factor fibre optic connector resembling a small SC connector. The LC connector has been standardised as FOCIS 10 in EIA/TIA-604-10. The ST/SC/LC connectors are illustrated in Figures 1.6, 1.7 and 1.8, respectively. We now move on the description of digital communication system.

1.4 Digital communications System

Certain advantages of the digital communication system and its underlying technologies makes its performance superior to analog systems. Thse include better signal-to-noise ratio (SNR)

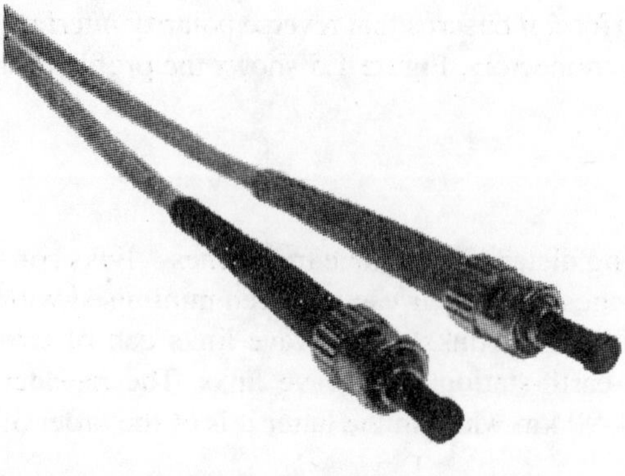

Figure 1.6 *The ST connector*

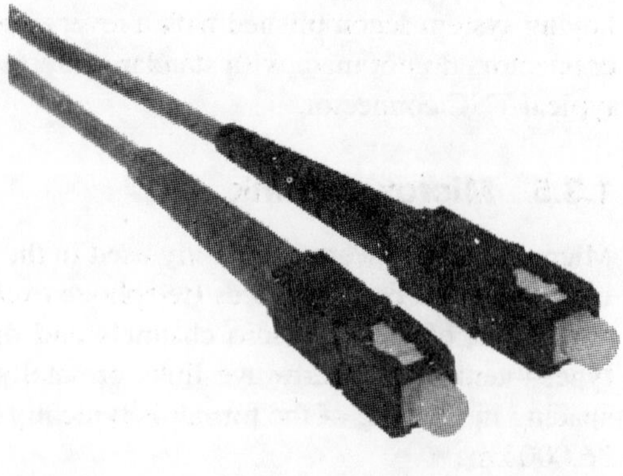

Figure 1.7 *The SC connector*

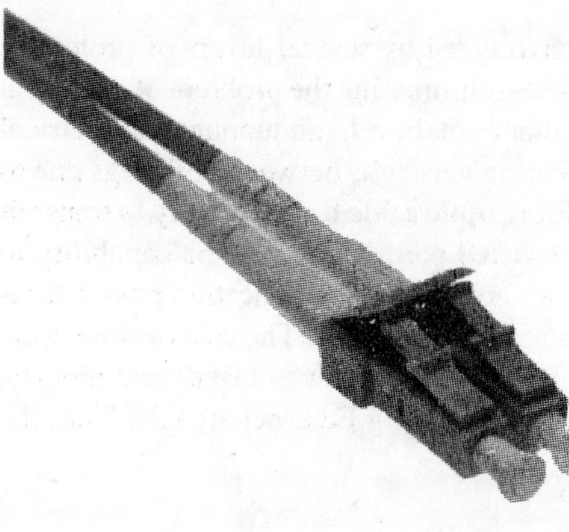

Figure 1.8 *The LC connector*

in some noisy channels, lesser power consumption and better efficiency. Digital systems are less sensitive to component aging. They can also be easily modified or reconfigured to meet future needs. The superb computational powers of the present-day central processing units (CPUs) and digital signal processors (DSPs) make digital communications more reliable, less expensive and more efficient compared to its analog counterpart. And with the present generation of encryption techniques, they are more secure. The digital transmission hierarchy can not only provide high bandwidth but can also integrate various types of communications, e.g. voice, file transfer, video, etc. into a single channel, often referred to as the 'bit pipe'. Advances made in transmission and as well as switching technology has made it possible for commercially viable technical solutions for meeting the requirement of high data rates and high speed switching. The digital processing system employed in digital communications offer the following advantages:

1. Affordable implementation of certain functions like nonlinear functions, linear phase filters and 2D filters which are too expensive or impractical to implement in the analog domain.
2. Insensitivity to component tolerances, ageing and temperature drift.
3. More predictable and repeatable system behaviour

4. Easy reprogrammability
5. Its size is independent of component values and allows high levels of integration.

We now briefly discuss the important concepts in digital processing of analog signals.

1.4.1 Digital signal processing

A signal is a function carrying information. Digital signal processing (DSP) is the application of mathematical operations on the digitally represented signals. Digital signals often represent signals from physical systems, which are in relation to physical time. DSP is an advanced technological forefront. The availability of several low cost and fast DSP chips, A/D converters and D/A converters have made many applications easily realisable in the digital domain. Lucent Technologies, Texas Instruments and Analog Devices are some of the major manufacturers of DSP chips today.

1.4.2 Limitations of digital communication system

Digital communication systems are not entirely without drawbacks. In some cases a signal may not lend itself to easy digital-to-analog conversion due to the large bandwidths involved. Digital systems are generally more expensive and more complicated to build and maintain than analog. There are certain applications where only analog solutions seem cost-effective. We can summarise the limitations of digital processing as applicable to digital communications as follows.

1. The signal bandwidth is limited by sampling rate.
2. Processors have limited dynamic range.
3. There is an additional noise component in digital system—the quantisation noise.
4. The performance is affected by round-off errors due to the finite register lengths.

Many of the above drawbacks have been overcome with the advances made in processor architecture, transmission hierarchy and transmission/switching techniques. New processing techniques like PDH, SDH, LDM and integration of different services put digital communication systems at an advantageous position as compared to analog communications. A detailed discussion on the above topics is beyond the scope of this book. The main reasons however for digital communications moving ahead of analog communications are:

1. the digital transmission hierarchy which provides high volume of bandwidth and integrates various means of communication, viz. voice, data, and video,
2. advantages in transmission and switching techniques that makes high data rate and high-speed switching possible.

1.5 The Concept of Bandwidth

It should be made clear at this level that all channels, be it analog or digital, have a capacity which limits the transmission of signals through it. The well-known *Shannon's theorem on channel capacity* states that *if the rate of information from a source does not exceed the capacity of the communication channel, then there exists a coding technique such that the information can be transmitted over the channel with an arbitrarily low frequency of errors despite the presence of noise.*[5] On the other hand, if the rate of information from the source

[5]Claude Elwood Shannon (1916–2001) founded the subject of Information Theory and proposed a linear schematic model of a communications system.

is greater than the channel capacity, the error rate of the received signal is enormous and detection of the original message is impossible. It can be easily established that the rate of information from the message source is a clear indication of the signal bandwidth—higher the rate, higher is the bandwidth of the signal. Thus, audio signals have a lesser bandwidth than video signals. Typical bandwidths of commonly encountered signals are given in Table 1.3.

Table 1.3 *Typical signal bandwidths*

Signal Type	Bandwidth in Hertz
ECG signal†	0.5 Hz–100 Hz
EEG signal	40 Hz–56 Hz
EMG signal	100 Hz-200 Hz
audio (speech)	4.0 kHz
fax message	10 kHz
data (computer)	300 Hz–1 MHz
music	16 kHz
video	6 MHz

†Courtesy: American Heart Association (AHA)

Claude Shannon, Father of Information Theory, Dies at 84

Murray Hill, N.J. (February 26, 2001) – Claude Elwood Shannon, the mathematician who laid the foundation of modern information theory while working at Bell Labs in the 1940s, died on Saturday. He was 84.

Shannon's theories are as relevant today as they were when he first formulated them. "It was truly visionary thinking," said Arun Netravali, president of Lucent Technologies' Bell Labs. "As if assuming that inexpensive, high-speed processing would come to pass, Shannon figured out the upper limits on communication rates. First in telephone channels, then in optical communications, and now in wireless, Shannon has had the utmost value in defining the engineering limits we face."

In 1948 Shannon published his landmark paper *A Mathematical Theory of Communication.* He begins this pioneering paper on information theory by observing that "the fundamental problem of communication is that of reproducing at one point either exactly or approximately a message selected at another point." He then proceeds to so thoroughly establish the foundations of information theory that his framework and terminology remain the standard even today.

Shannon's theory was an immediate success with communications engineers and stimulated the technology that led to today's Information Age.

*h*ttp://www.bell-labs.com/news/2001/february/26/1.html.

It has been made evident that a channel with more capacity can convey more messages simultaneously by means of a technique known as *multiplexing*.The optical fibre cable (OFC) and the microwave links have large capacity and can carry several simultaneous telephone signals; hence they are used in the telephone *trunk circuits*. We will discuss them in the next chapter.

1.6 Baseband Signals and the Need for Modulation

The band of frequencies of the signal delivered by the source or the input transducers are called baseband signals. The Fourier transforms of these analog signals occupy frequencies extending to (or close to) zero (dc); they are not shifted to some non-zero point on the frequency axis. For example, telephone quality speech occupies the band of frequencies from 300 Hz to 3.3 kHz. In television, the baseband is the video signal occupying 0 to 4.3 MHz. Although baseband signals can be transmitted as such (as in the case of telephony), it is quite often not the case. One of the inherent limitations of the baseband signal is that it is highly inefficient for electrical transmission over free space. As a rule, for effective radiation of an electromagnetic wave, the antenna length has to be at least $\frac{1}{10}$th its wavelength, which makes the length of baseband signal antenna impractically long.[6] Another limitation is that we cannot use the channel to carry multiple signals (multiplexing) if baseband transmission is resorted to. Hence for transmission purposes the baseband signal is used with a carrier signal of high frequency, wherein the baseband modulates (or varies) a characteristic of the carrier—signal amplitude in amplitude modulation (AM) or frequency in frequency modulation (FM), etc. This will be discussed in a subsequent chapter.

The transactions between the message source and the destination can take place in many ways. We now consider the possible transaction modes commonly employed in communication systems.

1.7 Modes of Transactions

The transaction between two points of information sources (P-to-P) can be done in three different ways. The simplest is the unidirectional mode of transmission from the 'talker' to the 'listener'. This mode of information transmission is referred to as SIMPLEX (in short SPX). Here the listener node has no right to talk back as in the case of radio or TV broadcasts. This kind of transaction is of limited use. The second mode of transaction is called half-duplex (HDX), where the listener too can talk, but not simultaneously with the talker. Normally, there will be a protocol (or merely a set of rules) to be adopted, by which they pass on information to and fro in a time-shared manner. An example of HDX mode is amateur radio broadcast and reception. The most sophisticated mode of transaction is the simultaneous transmission of information between two points, which is referred to as full-duplex (FDX). An example of FDX is the telephone system. These three modes are illustrated in Figure 1.9.

Appendix–MATLAB as a Programming Tool

In this appendix we attempt a brief introduction to MATLAB programming. MATLAB is considered to be the most suitable programming language for communication system modelling and simulation.

Introduction to MATLAB

MATLAB is an acronym for 'MATtrix LAB'. It is a high performance numerical computation and visualisation language. MATLAB is supported by almost all operating systems. MATLAB

[6] For a 300 Hz baseband audio signal, the wavelength $\lambda = c/f = \frac{3\times10^8}{300} = 1000$ km, an impractical height for an antenna.

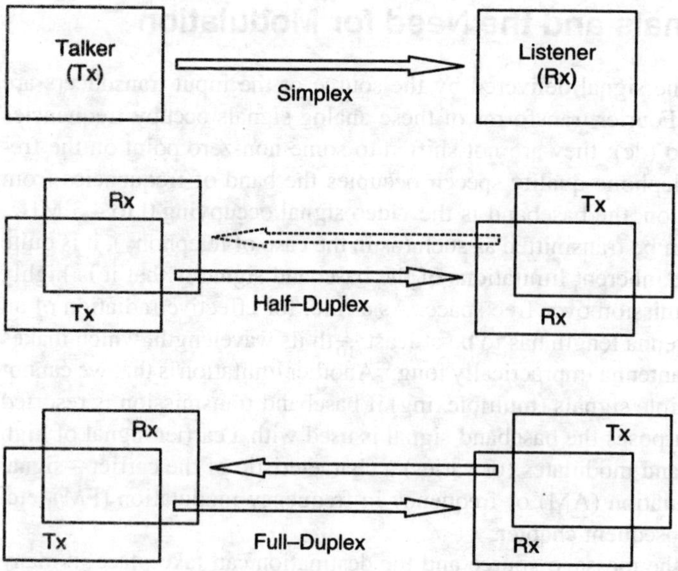

Figure 1.9 *Various modes of transaction*

Version-1 released in late 1970's was developed jointly by teams at University of New Mexico and Stanford University. Now, MATLAB is a registered trademark of Math Works, Inc., USA. Basically, MATLAB is an interpreter-based language. However compilers are available for later versions. The latest version available now is the 7.7 (MATLAB R2008b) released in September 2008. MATLAB, which was designed specifically for mathematics, lends itself beautifully to the needs of signal processing.

Basic features

The fundamental data element in MATLAB is an array and dimensioning is not necessary. Vectors ($1 \times N$ or $N \times 1$ arrays) are the next simpler data type. Integers are supported in short/long formats. Strings/structures are another data type. MATLAB is optimised for array/vector processing. Certain features of C/C++ (e.g. inheritance and operator overloading) is carried over. MATLAB has a very rich repertoire of built-in mathematical/logical functions. Users can add on to the list of such functions. Collections of special functions are available as toolboxes. Communications engineering students may find the following toolboxes most suitable for their course of study:

- Signal processing toolbox
- Control system toolbox
- Communications toolbox
- Simulink
- Neural network toolbox
- Fuzzy logic toolbox
- Signal processing blockset
- Communications blockset

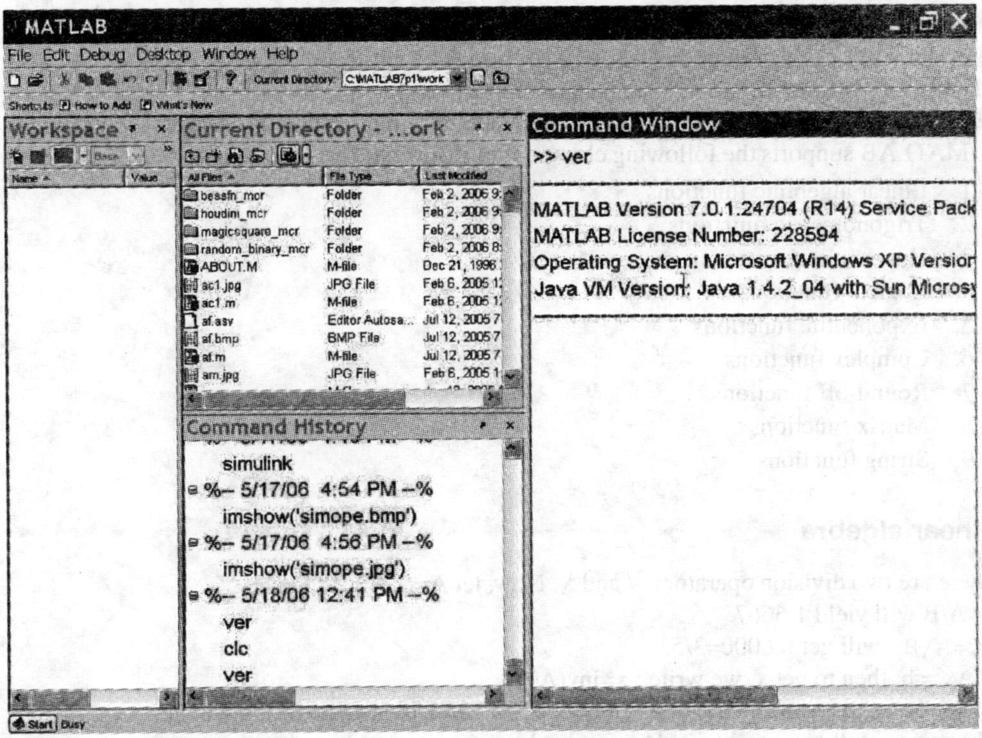

Figure 1.10 *Typical MATLAB opening window*

The typical MATLAB command window is illustrated in Figure 1.10.

Enhanced graphical features make programming and the displaying of results quite user friendly. External interface with C/Fortran is also possible by the world. There are several MATLAB importing '.MEX' files. Importing/exporting data from/to MS Excel/MS Word are possible. Any symbolic algebraic computations that can be done by *Mathematica, Maple and Macsyma* can be done with MATLAB symbolic toolbox. But the learning curve for MATLAB is shallow compared to them. MATLAB is the 'clones' available, some of them in the most popular and supported language among the technical computing community around open source regime. They are, *Octave for Linux under GNU*, *Maxima*, and *Scilab for Linux*.

Elements of MATLAB programming

To enter a 3 × 3 array into MATLAB workspace, we type the following:

```
a=[1 2 3;4 5 6;7 8 9]
```

This will result in:

```
a =

    1     2     3
    4     5     6
    7     8     9
```

If we add a semicolon(;) at the end of command, the values in the matrix **a** will not be echoed. Now, let A=[1 2;3 4]; Then b=A+2; will get b= [3 4;5 6]. If a=[1 2; 1 0]; and b=[1 -1; 0 2]; Then c=2*a will get [2 4;2 0] in c. (scalar multiplication). c=a*b will get [1 3;1 -1]. (matrix multiplication). Now, c=a.*b will result in elementwise multiplication and get [1 -2;0 0] in c.

MATLAB supports the following classes of built-in (library) functions:

1. Linear algebraic functions
2. Trigonometric functions
3. Hyperbolic functions
4. Logical functions
5. Exponential functions
6. Complex functions
7. Round-off functions
8. Matrix functions
9. String functions

Linear algebra

There are two division operators –/ and \. Now, let A=5; B=3; then:
C=A/B will yield 1.6667.
`C=A\B` will get 0.6000=3/5.
If Ax = b; then to get x, we write : x=inv(A)*b.
Example:- A=[6 -4 8;-5 -3 7;14 9 -5]; b=[112 75 -67]'; then x=inv(A)*b will get: x=[2 -5 10]'.
The same result can be obtained by `x= A\b`.

Operator overloading

If A is defined as above, then we define

```
A^2 = A*A = [168    60    -20;83     92    -96;-31   -128    200].
{This  is true  matrix  multiplication.}

A.^2 = [36     16     64;25      9     49;196      81     25].
{This  is referred to as elementwise multiplication.}
```

Some of the other important linear algebraic functions are illustrated below.

- `det(a)` will get the determinant of a square matrix a.
- `eig(a)` will get the eigen values of a.
- `[U  R]=eig(a)` calculate the eigen values and eigen vectors of a.
  Example:- `a = [3 -2 0;2 -2 0;0 1 1]`; then `eig(a)` gets

```
ans =
      1
     -1
      2
```

- `eye(5)` gets a 5 × 5 'Identity' matrix.
- `ones(1,5)` gets [1 1 1 1 1].
- `zeros(3,1)` gets [0 0 0]'.
- `size(eye(3,1))` gets [3 1].

- `length([1 0 7 -4])` gets 4.
- `sum([1:5])` gets 15.
- `prod([1:5])` gets 5!=120.
- If a=[2 -3 0 5 7], find(a>0) gets [1 4 5].
- a(a<0) gets -3.
- a(find(a~=0)) or a(a~=0) gets [2 -3 5 7].
- a~=0 gets [1 1 0 1 1].
- Let A=[6 -4 8;-5 -3 7;14 9 -5]; b=[1 2 3]; then C=[A;b] will get [6 -4 8;-5 -3 7;14 9 -5;1 2 3] in C.% A row is appended. \% is used to comment a line. MATLAB interpreter will ignore whatever follows a \% symbol.
  Also C=[A b'] gets [6 -4 8 1;-5 -3 7 2;14 9 -5 3] in C. % A column is appended.
  Now C(2,:)=[]; deletes the second row of C.
  Also C(:,3)=[]; deletes the third column of C.
- `rand(m,n)` % returns an $m \times n$ matrix of uniformly distributed random numbers, between 0 and 1.
- `randn(m,n)` % returns $m \times n$ matrix of normally (Gaussian) distributed random numbers with zero mean and unity variance.
- `diag(v)` % generates a diagonal matrix with the vector v on the diagonal.
- `diag(A)` % extracts the diagonal of matrix A as a column vector.
- `diag(A,1)` %extracts the first upper off-diagonal vector of A.
- `rot90(A);` % rotates A by 90 degrees.

```
A= [6 -4 8;-5 -3 7;14 9 -5];
rot90(A)
ans=
    8  7 -5
   -4 -3  9
    6 -5 14
```

- `fliplr(A);` flips a matrix from left to right.
- `flipud(A);` flips a matrix from up to down.
- `tril(A);` extracts the lower triangular part of a matrix.
- `triu(A);` extracts the upper triangular part of a matrix.
- `reshape(B,4,2);` If B is 2×4, it is reshaped as 4×2.

Array operators

- .* element-by-element multiplication.
- ./ element-by-element left-division.
- .\ element-by-element right-division.
- .^ element-by-element exponentiation.
- .' non-conjugated transpose.

Relational operators

- If x=[1 5 3 7] and y=[0 2 8 7]; then:
  < less than; `K=x<y` gets K=[0 0 1 0].
- <= less than or equal to; `K=(x<=y)` gets K=[0 0 1 1];
- > greater than; `K=x>y` gets K=[1 1 0 0];
- >= greater than or equal to; `K=(x>=y)` gets K=[1 1 0 1];
- == equal to; K=(x==y) gets K=[0 0 0 1];
- ~= not equal to; K=(x~=y) gets K=[1 1 1 0];

Logical functions

MATLAB supports the following logical functions/operators.

- If x=[0 5 3 2] and y=[0 3 0 5];then:
  & logical AND; K=x&y=[0 1 0 1];
- | logical OR; K=x | y=[0 1 1 1];
- ~ logical complement (NOT); K=~x=[1 0 0 0];
- `xor` exclusive OR; K=xor(x,y)=[0 0 1 0];

Built-in logical functions

- `all;` If a=[-1 -8 -5]; then all(a<0) returns 1.
- `any;` If b=[0 5 7]; then any(b>5) returns 1.
- `exist;` if matrix A exist in workspace, then exist('A') returns 1.
- `isempty;` If b=[]; then isempty(b) returns 1.
- `isinf;` returns 1, iff argument is infinity.
- `isfinite;` returns 1, iff argument is finite.
- `isnan;` returns 1, iff argument is not a number.
- `find;` If a=[-1 7 -5]; then find(a)=[0 1 0].

Exponential functions

Examples of exponential functions are given below:

- If a=[0 1]; then `exp(a)=[1 e];`
- If a=[1 e]; then `log(a)=[0 1];`
- If a=[10 100]; then `log10(a)=[1 2];`
- If a=[25 49]; then `sqrt(a)=[5 7];`

Complex functions

MATLAB has the following functions that operate on complex entities.

- `abs;` % abs(3+4i) returns 5 and abs(-3) returns 3.
- `angle;` % angle(3+4i) gets 0.9273 rad.
- `conj;` % conj(3+4i) returns 3-4i.
- `imag;` % imag(8+7i) returns 7.
- `real;` % real(8+7i) returns 8.

Round-off functions

- `fix;` % rounds towards zero. fix([-2.33 2.88]) =[-2 2].
- `floor;` % rounds towards -infinity.
  `floor([-2.33 2.66])=[-3 2].`
- `ceil;` % rounds towards +infinity.
  `ceil([-2.33 2.88])=[-2 3].`
- `round;` % rounds towards nearest integer.
  `round([-2.33 2.66])=[-2 3].`
- `rem;` % rem(a,b)=a-fix(a./b).
- `sign;` % signum function, `sign([-2.3 0 2.6])=[-1 0 1].`

Matrix functions

These are special functions that accept a matrix as an argument, as a whole.

- `expm(A);` % finds the exponential of the matrix A, exp(A).
- `logm(A);` % finds log(A) such that A=exp(log(A)).
- `sqrt(A);` % finds $\sqrt{A}$.

String functions

MATLAB has a rich collection of functions that operate on strings.

- `char;` creates character arrays using automatic padding. Also converts ASCII numeric values to character arrays.
  Example:- `howdy=char('Hi','Hello','Namaste');` results in:

```
howdy =
Hi
Hello
Namaste
```

  `char(65)` returns A.
- `abs;` % converts characters to their ASCII numerical values. abs('A') returns 65.
- `blanks(n);` % creates n blank characters.
- `deblank;` % removes the trailing blanks from a string.
- `eval;` % executes the string as a command, if they contain valid MATLAB commands. `eval('y=5*sin(x)')` does the same thing as y=5*sin(x).
- `findstr;` finds specified substring.

```
s = 'How much wood would a woodchuck chuck?';
        findstr(s,'a')       returns  21
        findstr('a',s)       returns  21
        findstr(s,'wood')    returns  [10 23]
        findstr(s,'Wood')    returns  []
        findstr(s,' ')       returns  [4 9 14 20 22 32]
```

- `int2str;` converts integers to string.
- `ischar;` true (=1) for character arrays.
- `isletter;` true (=1) for alphabetical characters.
- `isstr;` true if argument is a string; isstr('MATLAB') returns 1.
- `lower;` upper case to lower case conversion.
- `upper;` lower case to upper case conversion.
- `mat2str;` converts a matrix to string.
- `num2str;` converts numbers to strings. (similar to int2str)
- `strcmp;` compares two strings, returns 1 if same.

```
s='malayalam';
>> p=fliplr(s);
>> strcmp(s,p)
ans =
      1
```

- `strncmp;` compares first n characters in given strings. strncmp(s,p,9)=1, but strncmp(s,p,10)=0.
- `strcat;` concatenates strings ignoring trailing blanks.

```
P=strcat('It rains cats', ' and dogs.') returns
P=It rains cats and dogs.
```

- `strvcat;` similar to char, but ignores empty strings.

```
P=strvcat('It rains cat', '', ' and dogs.') returns:
P =
It rains cats
and dogs.
```

Plotting with MATLAB

MATLAB has a number of plotting commands.

- `plot(x,y,'r+');`
- `fplot('function',[xmin xmax]);` function plotter.
- `ezplot('function',[xmin xmax]);` easy function plotter.
- `stem(x,y);` plots a discrete signal.
- `ezpolar('function', [0 2*pi]);` polar version of ezplot.
- `plot3(x,y,z);` 3D plot.
- `ezplot3('f1','f2','f3', [0 2*pi]);` 3-D plotter.
- `contour(X,Y,Z);`
- `ezcontour(Z);`
- `ezcontourf(Z);`
- `surf(X,Y,Z);`
- `ezsurf(Z);`
- `ezsurfc(z,[x,y]);`

Specialised plotting

- `mesh(x,y,z);`
- `meshc(x,y,z);` same as mesh, but draws a contour plot under the surface.
- `meshz(x,y,z);` same as mesh but draws a series of vertical lines under the surface.
- `ezmesh(z);` easy mesh.
- `ezmeshc(z);` easy meshc.
- `waterfall(x,y,z);` same as mesh but draws mesh lines in one direction only.
- `bar(n,x);` creates bar graph.
- `[x,y]=meshgrid(min:max);` creates the matrices x and y from the vector `[min:max]` which are the limits of x and y.

The plot of the jointly Gaussian random variable obtained by the MATLAB `ezmeshc` command and the corresponding code are shown in Figure 1.11.

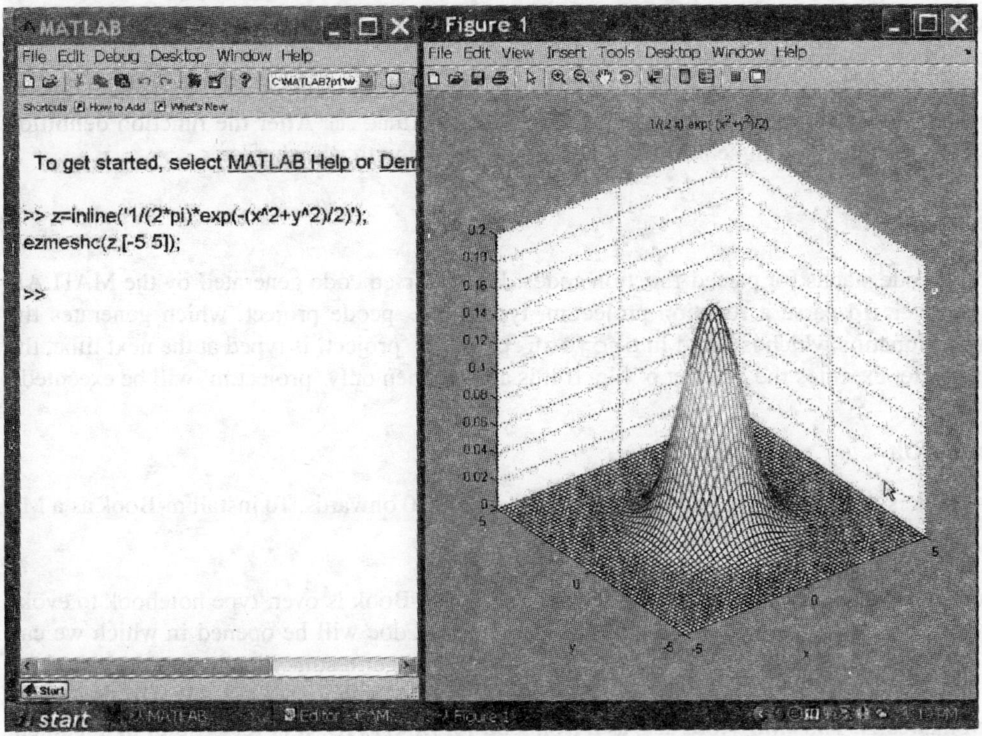

Figure 1.11 *Plot of a jointly Gaussian random variable*

Language-specific features

In MATLAB, elementwise array operators are preceded by a period (.). A line can be continued by putting three consecutive periods (...) at the end of the line. Comma is the delimiter for elements of a vector, array or different arrays/vectors. Semicolon is the statement delimiter. Output will be echoed only if the semicolon is absent. The array index starts at 1. Zero cannot be the starting index. A comment can be entered by typing a % character, and the MATLAB interpreter ignores whatever succeeds it.

m–files

M-files enable us to write and execute a number of MATLAB commands at a single stretch. There are two types of m–files.

Script m–files
A script m–file is equivalent to a macro definition. Parameter passing is not intended but can share parameters already in workspace. Less efficient compared to function m–file.

Function m–files
The function m–file is equivalent to a function call. Nesting of functions is permitted. The comment lines given after the function definition line appears when help on that function is requested.

Function definition line

The general form of a function definition line is:

```
function [output variabale]=function_name(input variables);
```

The above function has to be saved as `function_name.m`. After the function definition line any number of comment lines can be given starting with a % character.

P–Code

The P-code stands for parsed-function codes. It is a parsed code generated by the MATLAB interpreter. To parse a function project.m, type: >> pcode project, which generates the parsed function which is saved in `project.p`. When 'project' is typed at the next time, the interpreter executes the 'project.p' file. If it is absent, then only 'project.m' will be executed.

m–Book

m–Book is a feature supported by MATLAB version 6.0 onwards. To install m-Book as a MS Word overlay, type:

```
>> notebook  -setup;
```

and follow the instructions. Once the installation of m-Book is over, type notebook to evoke the m-Book. A word document with file name mbook.doc will be opened in which we can type MATLAB commands followed by 'Cntl-Enter' to get results identical to that obtained at the workspace. If we open the mbook.doc file from MS Word, MATLAB will be opened automatically. Thus MS Word acts as a front end for MATLAB. It is very useful in document preparation.

MATLAB profiler

To evaluate the performance of functions, MATLAB version 5.0 onwards provides a tool called profiler. The profiler keeps track of the time spend on each line of the function (in units of 0.01 seconds) as a function is executed. The profile report shows how much time was spent on a particular line. The profile output can also be seen graphically as a Pareto plot. The following commands will invoke the MATLAB profiler.

```
profile on
File_name
profile report
profile plot
profile off
```

Control flow

The following control constructs are available in MATLAB:

- `for-end` loop. For loops can be nested.
- `while-end` loop. Can be nested, but every while loop must have a matching end.
- `if-elseif-else-end` statement. Can be nested.
- `switch-case-otherwise-end` statement.
- `break;` for terminating the `for` and `while` loops prematurely.

- `error.` The command error ('message') inside a function will abort the function and display the error message 'message'.
- `return.` Returns the control to the invoking function.

Interactive input

There are mainly four commands for interactive inputs.

Input

```
L=input('Length of the vector ')
Length of the vector 5
L =
     5
yn=input('More Simulations? (y/n) ','s')
More Simulations? (y/n) y
yn =
     y
```

Keyboard

On execution of the keyboard command, the command window prompt changes to `>>k`. The user can examine the status of the program variables up to that level by typing the appropriate variable name. After finishing, type return and press the return key to get back to the execution of the remaining part of the program.

Menu

The menu command creates an on-screen menu.

```
p=menu('plot','circle','polar','cartesian');
```

On pressing the cartesian option:

```
p=3
```

Pause

The `pause` temporarily halts the execution of the program. Pressing any key resumes execution. `Pause(n);` halts for n seconds and then resumes.

Special MATLAB functions

There are several special MATLAB functions that engineering students will find useful. They are included in Control System Toolbox and Signal Processing Toolbox. Some general purpose functions are listed below.

Statistical functions

- `primes(N);` lists out all primes less than or equal to N.
- `isprime(N);` checks if N is a prime.
- `factor(N);` factors N into prime factors.
- `std(A);` computes the standard deviation of elements of vector A. If A is a matrix, then it computes the standard deviations of each column.
- `mean(A);` computes the mean of elements of vector A. If A is a matrix, then it computes the mean of each column.
- `var(A);` similar to std(.), but computes variance.
- `bar(A);` plots a bargraph using data from vector A.

Signal processing functions

Users will find the functions listed below useful for signal processing applications.

- `modulate(.)`; modulator function
- `conv(X,H)`; obtains the linear convolution of sequences X and H
- `xcorr(X,H)`; obtains the cross-correlation between sequences X and H. `xcorr(X,X)` computes the autocorrelation of sequence X
- `fft(X,N)`; computes the N point fast Fourier transform (fft) of sequence X
- `ifft(X)`; computes the inverse fft
- `dftmtx(N)`; computes $N \times N$ DFT of matrix D_N
- `fftfilt(h,x)`; overlap-add method implementation of linear convolution
- `zplane(zeros,poles)`; z-plane plot of poles and zeros
- `zp2sos(zeros,poles,k)`; zero-poles to second order section conversion
- `[r,p,k]=residue(num,den)`; partial fraction expansion in s-domain
- `[r,p,k]=residuez(num,den)`; partial fraction expansion in z-domain
- `[h,t]=impz(num,den)`; inverse z-transform
- `filter(num,den,X)`; response of the LTI filter with numerator polynomial coefficients 'num', and denominator coefficients 'den' and the input sequence X
- `[N,Wn]=buttord(Wp,Ws,Rp,Rs,'s')`; computes the lowest order N of a Butterworth analog transfer function meeting the specifications given
- `[num,den]=butter(N,Wn,'s')`; analog Butterworth filter
- `[N,Wn]=cheby1ord(Wp,Ws,Rp,Rs,'s')`; Type 1 Chebyshev filter order N
- `[num,den]=cheby1(N,Rp,Wn,'s')`; Chebyshev analog lowpass filter transfer function
- `[N,Wn]=ellipord(Wp,Ws,Rp,Rs,'s')`; Order N of elliptical filter
- `[num,den]=ellip(N,Rp,Rs,Wn,'s')`; Elliptical analog lowpass transfer function
- `[num,den]=besself(N,Wn)`; used to design Nth order Bessel filter
- `[b,a]=lp2lp(num,den)`; lowpass to lowpass conversion
- `[b,a]=lp2hp(num,den)`; lowpass to highpass conversion
- `[b,a]=lp2bp(num,den,Wp1,Wp2)`; lowpass to bandpass conversion
- `[num,den]=bilinear(b,a,Ws)`; bilinear transformation to get a digital filter equivalent
- `fvtool(num,den)`; filter visualisation tool, available only in MATLAB ver 7.0.
- `fdatool(num,den)`; filter design and analysis tool
- `sptool(.)`; Signal processing tool. `sptool` is a graphical user interface (GUI). `sptool` opens the signal processing tool window which allows you to import, analyse and manipulate signals, filters, and spectra.

The profiles of the *Filter Visualisation Tool (fvtool)* and *Filter Design and Analysis Tool (fdatool)* windows are shown in Figures 1.12 and 1.13.

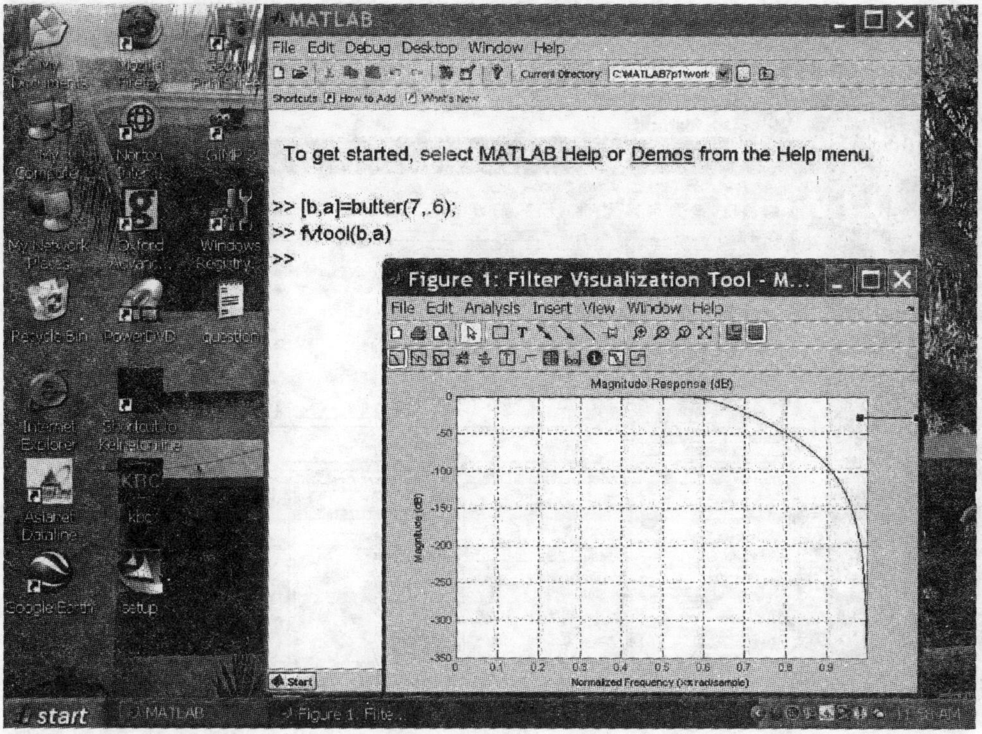

Figure 1.12 *Filter visualisation tool window*

Control system functions

There are several special functions in the control system toolbox useful for control system design.

- `r=roots(p);` roots of the polynomial coefficients p in descending powers of s
- `p=poly(r);` reassemble polynomial from roots
- `printsys(num,den);` displays the transfer function $H(s)$
- `impulse(num,den);` impulse response
- `step(num,den);` step response
- `feedback(sys1,sys2);` overall response of two systems, `sys1` and `sys2`, the latter is in the feedback loop
- `[z,p,k]=tf2zp(num,den);` transfer function to zero-pole conversion
- `[num,den]=zp2tf(Z,P,K);` zero-pole to transfer function conversion
- `[num,den]=ss2tf(A,B,C,D);` state-space to transfer function conversion
- `[A,B,C,D]=tf2ss(num,den);` transfer function to state-space conversion
- `bode(num,den);` Bode plot of LTI system
- `[r,k]=rlocus(num,den);` Root locus of LTI system
- `rlocfind(num,den);` rlocfind follows the rlocus function
- `[re,im]=nyquist(num,den);` Nyquist plot of LTI system
- `[mag,phase,w]=nichols(num,den,w);` Nichols chart

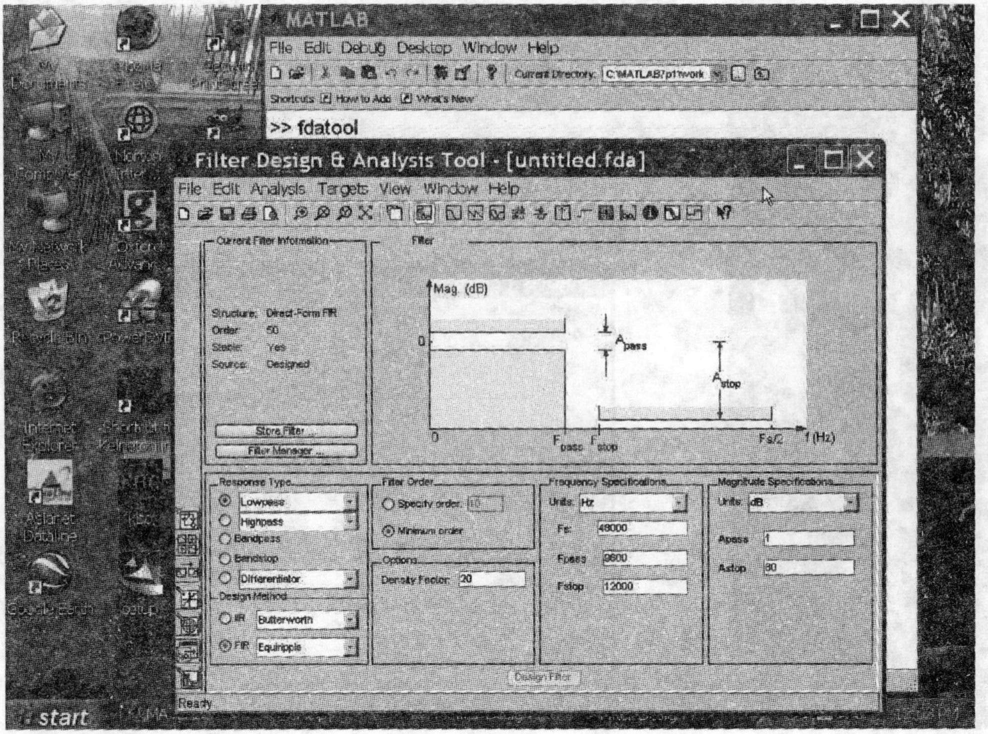

Figure 1.13 *Filter design and analysis tool window*

- `freqs(num,den)`; frequency response (continuous time) of a system
- `freqz(num,den)`; frequency response (discrete time) of a system
- `[out,state]=lsim(A,B,C,D,u,t,x0)`; output and state response of a LTI system.

Simulink

Simulink is a block level system simulation tool available in MATLAB which is quite user friendly as it is *GUI* based. Simulink can be invoked from MATLAB command window by typing `simulink`. Typical Simulink opening window is shown in Figure 1.14.

Generating executable files in MATLAB

It is possible to generate executable files from MATLAB script/function files. You have to use the MATLAB compiler available with the full MATLAB suite. Try the documentation: `>> doc complier;` for details on MATLAB complier. If you have, for example, the `magicsquare.m` file in the `/MATLAB7p1/work` subdirectory, the compiler can be evoked by typing: `>> mcc -mv magicsquare.m;` Follow the instructions and you will get the MS DOS executable `magicsquare.exe` in the work subdirectory, after compilation. We need not have to open MATLAB once the `magicsquare.exe` file is generated: it can be run like any other executable file.

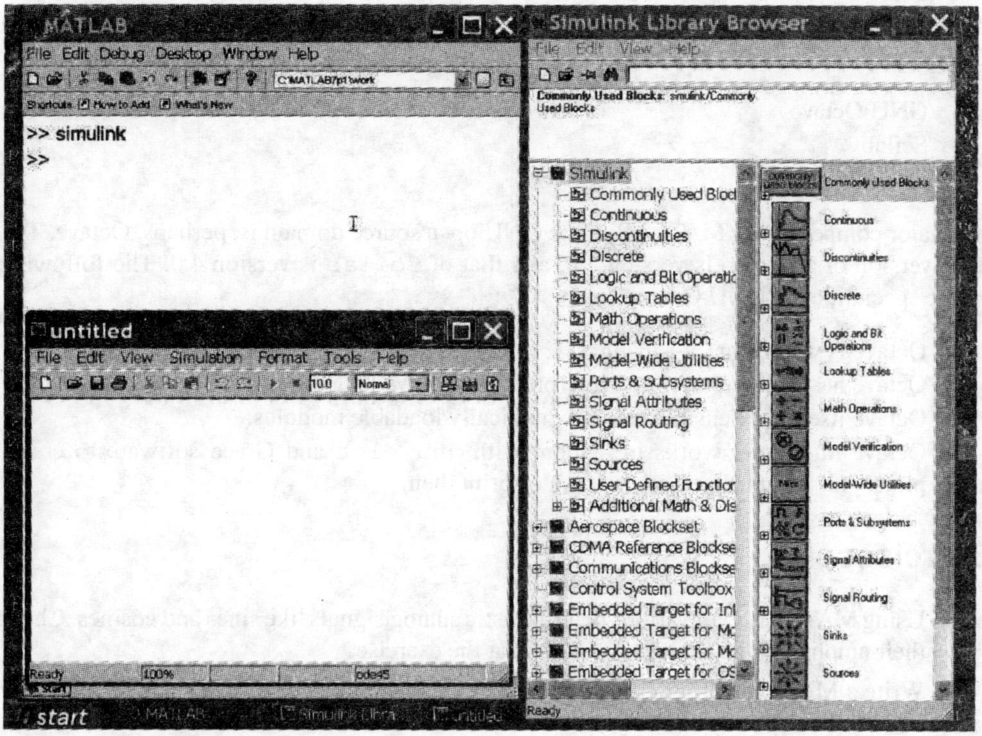

Figure 1.14 *Typical Simulink opening window*

MATLAB help

To get online help on MATLAB topics, the following can be employed.

- help *topic*. (for on-line help on 'topic.m') This command will list out all the *comments* (lines starting with a "%" character) given immediately after the function definition line.
- `helpwin`. (for listing out topics).
- doc *topic*. (for html documentation) This is the most elaborate online help available in MATLAB.
- lookfor *topic*. (a sort of keyword search, takes more time than the other two).

MATLAB clones

Several MATLAB-like programs are now available, some of them are commercial software tools and some of them are available in the open software domain. Examples of commercial equivalents of MATLAB are:

- Mathematica
- Mathcad
- Maple
- Macsyma

MATLAB equivalents in the open source domain available under the general public license (GPL) are:

- GNU Octave
- Scilab
- Maxima

The major competitor for MATLAB in the GNU open source domain is, perhaps, Octave. The latest version of `Octave` is version 2.0, and that of `Scilab` is version 4.0. The following are the specialities of GNU Octave:

- Octave is written in C++ using STL libraries.
- Octave has an interpreter that interprets the Octave language.
- Octave itself is extensible using dynamically loadable modules.
- Octave interpreter works in tandem with `gnu plot` and Grace software to create plots, graphs and charts and to save or print them.[7]

Exercises

1. Using MATLAB, plot various deterministic analog signals like sines and cosines. Change their amplitude and frequency and repeat the exercise.
2. Write a MATLAB program to generate an exponential waveform for $0 \leq t \leq 2$. Now modify the code so that N cycles of the waveform is generated and plotted.
3. Write a MATLAB program to generate 'Lissajous figures'. Vary the vertical and horizontal signal frequencies and observe the variation in shape.
4. Generate a square wave as a sum of several sinusoids. Write a MATLAB m–file that will do the job and vary the number of harmonics and see the variation.
5. Plot a 1000 Hz sine wave and also plot its delayed version where the delay is $\pm\pi$. Now sum the two and plot. What do you observe?
6. Generate and plot a random signal using MATLAB (Clue: Use rand() and randn() functions).
7. Simulate ECG signals using MATLAB.
8. Plot the sampling function $Sa(t) = \frac{\sin(t)}{t}$ using MATLAB. Also plot $\left[\frac{\sin(t)}{t}\right]^2$ and $\frac{\sin(t^2)}{t^2}$.
9. For a 3 V half-wave rectified sine wave, obtain the Fourier series coefficients.
10. Determine the spectrum of the above wave using the fft() function in MATLAB.
11. Obtain the spectrum for a 3 V, 0.5 s rectangular pulse using the fft() function in MATLAB. Compare it with the theoretical spectrum.
12. Sketch, using MATLAB, a periodic function given by $f(x) = |\sin(x)|$ for $-\pi < x < \pi$. Determine whether it is an odd or even function.
13. The signal $x(t)$ is given by: $x(t) = \sin c(100\,t)\cos(2\pi \times 200\,t)$. Plot the function and its magnitude spectrum using MATLAB.

[7] See *GNU Octave Reference Manual* by John W. Eaton, 1997.

14. Plot the sinc $(x) = \frac{\sin(\pi\, x)}{\pi\, x}$ function for x in steps of 0.1 over the range $-2.2 \leq x \leq 2.2$.
15. One cycle of a periodic waveform is given by: $f(x) = \cos x$, for $0 < x < \pi$, and $-\cos\, x$, for $-\pi < x < 0$. Obtain the average value (dc) of the signal.
16. Simulate amplitude shift keying: when the data bit is '1', send $A_1 = 5\sin(\omega_a\, t)$ and when it is '0', send $A_0 = 2\sin(\omega_a\, t)$. Take $f_a = 1$ kHz.
17. Determine the Fourier series expansion of the signal

$$x(t) = |\cos\, 2\pi f_0\, t| \quad \text{(full-wave rectifier output)}$$

18. Let $x(t)$ be an arbitrary signal and define $x_1(t) = \sum_{n=-\infty}^{\infty} x(t - nT_0)$.

    (a) Show that $x_1(t)$ is periodic.

    (b) How can we write $x_1(t)$ in terms of $x(t)$ and $\sum_{n=-\infty}^{\infty} \delta(t - nT_0)$?
19. Use the convolution theorem to show that

$$\text{sinc}(t) * \sin c(t) = \sin c(t).$$

20. Find the output of an LTI system with impulse response $h(t)$ when driven by the input signal $x(t)$ when:

$$h(t) = \delta(t) + \delta^{'}(t).$$

21. Use the MATLAB function 'modulate()' to generate and plot AMDSB-TC (A3E), FM, PM, PWM and PPM waveforms.
22. The impulse response of a system is

$$h(t) = \begin{cases} t^3\, e^{-t^2} & 0 < t \\ 0 & t < 0. \end{cases}$$

Use the conv(.) function in MATLAB to find the response of the system to the pulse

$$x(t) = \begin{cases} A & 0 < t < T \\ 0 & \text{elsewhere,} \end{cases}$$

where A and T are real positive constants.

Chapter 2

Telephone Systems

In this chapter we will discuss in detail the principles of telephone systems, their working, and the recent trends in technology. The word *telephone* is derived from the Greek words *tele* meaning *far*, and *phone* meaning *sound*. Telephony thus involves the conversion of sound signals into audio-frequency analog electrical signals, which can be then be transmitted over an electric transmission system and reconverted to sound pressure signals at the receiving end. The telephone system is the simplest possible baseband analog communication system. As we have seen in the previous chapter, telephone systems were the one of the foremost means of analog audio signal transmission over a pair of copper wires (also known as cable pairs) dating back to the second half of the nineteenth century. And they remain important even today, connecting people over different continents around the globe. Telephones have both civilian and military applications, and the technologies supporting them have been kept updated, making them more and more cost-effective. There are over one billion land phones and two billion mobile phones in the world now, according to the IEEE. Modern day telephones integrate voice and data transmission on a single pair of copper wires using MODEMS and customer premises equipment (CPE).[1] The key innovations in telephony have been circuit switching, digitisation, separation of call control (signalling) from voice transfer, fibre-optic links, and service integration (computer–telephony integration or CTI). In the first-generation telephone systems, both signalling and the voice transmission were analog. By the year 1988, voice transmission between switches went digital along with signalling and control of switches. The subscriber's local loop remains both analog and digital even today.

2.1 Public Switched Telephone Network

A simplified block diagram of a public switched telephone network (PSTN) system and the basic telephone system are given in Figures 2.1 and 2.2 respectively. The equipment can be segregated into two: those at the customer premises and those at the central office (in common terms, the telephone exchange).[2] The customer premises equipment is mainly the telephone handset, which contains the receiver, the transmitter, the ringer and the dialler. The telephone

[1]A MODEM is a MOdulator–DEModulator circuit which connects the computer data line to the plain old telephone system (POTS).

[2]Central office (CO): A common carrier switching centre in which trunks and loops are terminated and switched. Synonyms are exchange, local central office local exchange, local office, switching centre, switching exchange and telephone exchange.

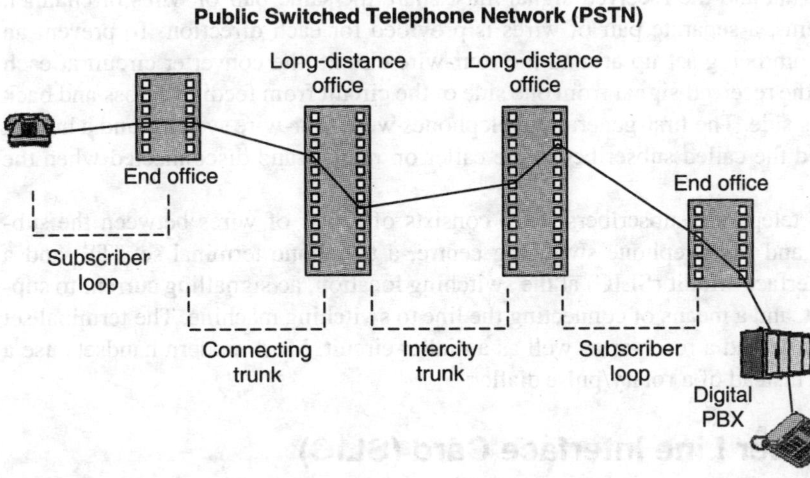

Figure 2.1 *Public switched telephone network*

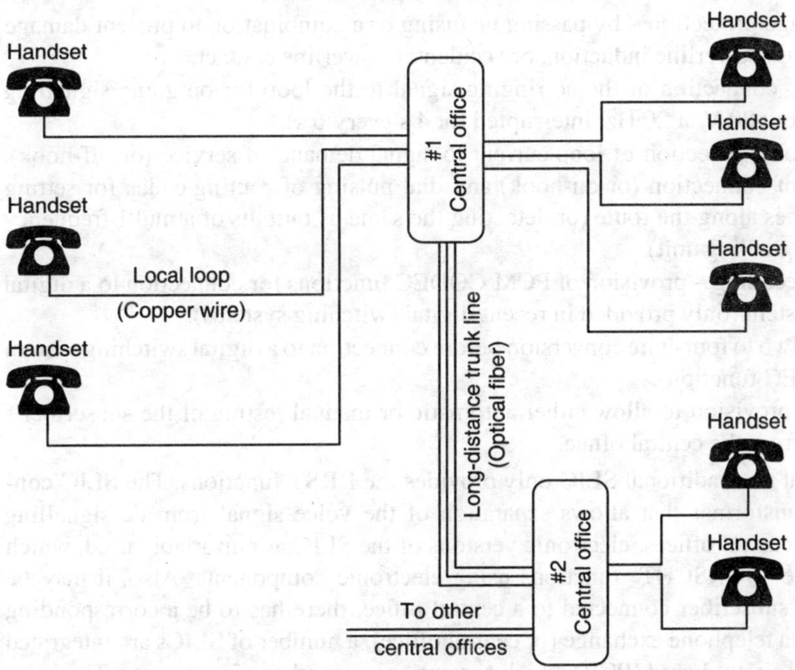

Figure 2.2 *The basic telephone system*

being a full-duplex (FDX) equipment, there is a need to provide-two way channels, one for each direction. This results in uneconomic duplication of facilities. In two-wire transmission, the transmitted signal and the received signal must share the same pair of wires or channel. In four-wire systems, a separate pair of wires is provided for each direction. To prevent an oscillation loop from being set up around the four-wire circuit, the converter circuit at each end must prevent the received signal from one side of the circuit from feeding across and back on the transmitting side. The first-generation telephones were four-wire systems and a human operator connected the called subscriber to the caller on request and disconnected when the call got over.

The traditional telephone subscribers' loop consists of a pair of wires between the subscriber's location and the telephone switching centre, a telephone terminal set (TS) and a subscriber line interface circuit (SLIC) at the switching location, ac signalling current to supply battery current, and a means of connecting the line to switching machine. The terminal set contains a transmitter and a receiver as well as a dialler circuit. Most modern handsets use a touch-tone dialler instead of a rotary/pulse dialler.

2.2 Subscriber Line Interface Card (SLIC)

The several functions of the SLIC, commonly called 'BORSCHT' as enumerated Below.[3]

- B: Battery – a 48 V central office battery supplies the dc loop current (about 40 mA) to energise the voice transmitter and the outgoing signal circuit.
- O: Overvoltage protection – by passing or fusing or a combination to prevent damage from lightning, powerline induction, or accidental powerline contact.
- R: Ringing – connection of the ac ringing signal to the loop for outgoing signalling (typically about 80 V at 20 Hz, interrupted for 4 s every 6 s).
- S: Supervision – detection of loop current to signal demand of service (or off-hook), termination of connection (or on-hook) and dial pulsing of routing codes for setting up the switches along the route (or detecting the same as outputs of a multi-frequency touch tone signalling unit).
- C: Coding/decoding – provision of PCM CODEC functions for connection to a digital switching system (only provided in recent digital switching systems).
- H: Hybrid – two to four-wire conversion before connection to a digital switching system (as for CODEC function).
- T: Testing – provision to allow either automatic or manual testing of the subscriber's loop circuit from the central office.

It may be noted that the traditional SLIC only provides the BRST functions. The SLIC contains a bridging transformer that allows separation of the voice signal from dc signalling current. In modern central offices, electronic versions of the SLIC are invariably used, which fully implement the 'BORSCHT' functions using electronic components. Also, it may be noted that for each subscriber connected to a central office, there has to be a corresponding SLIC. Normally, in a telephone exchange (or central office), a number of SLICs are integrated into a single printed circuit board (PCB) which can support a number of customers. These are called *line cards.* Quite often, working spare PCBs are made available in the central office so that they can be easily and quickly swapped with faulty ones.

[3] 'BORSCHT' means *red soup* in German.

2.3 Call Establishment and Release

The sequences of steps taking place during a call establishment are:

1. The caller lifts up the handset, which closes the hook switch to connect the handset to the line.
2. Operating the dialler, he/she sends several trains of 1 to 10 pulses at a rate of about 10 pulses per second to select the proper switch control in the switching machine.
3. If the called line indicates a 'busy' condition, a special busy tone is sent back to the caller on the voice circuit.
4. If not, the called party's loop is energised to sent bursts of ac to ring the bell and a ring tone is sent back on the voice circuit to let the caller know that the ringing is proceeding.
5. When the called party lifts the receiver, the hook switch closes the dc path in the called loop to disconnect the ringing signal.
6. Once the conversation is over, the calling subscriber goes on-hook and the connection is terminated by the release of the switching equipment.

2.3.1 Different tones used in telephony

They can be classified into two: dial signalling and informational signalling. Details of both are discussed in the following paragraphs.

Dial signalling
Dial signalling is the transmission of the dialling signals to the central office (telephone exchange) so that the calling subscriber is connected to the called subscriber. Dial signalling can be done either by means of pulse dialling (in the case of old pulse diallers) or by tone dialling (as in the case of modern touch-tone diallers). Diallers are discussed in greater detail later.

Informational signalling tones
This category includes the dial tone, ringing tone, busy tone, fast busy tone, ring-back tone, and number unobtainable tone. The frequencies used for them are tabulated in Table 2.1.

Table 2.1 *Telephone event tones*

Event	Frequency in hertz		Duration in seconds	
	Low	High	On	Off
Dial tone	350	440	Continuous	
Busy tone	480	620	0.5	0.5
Fast busy tone	480	620	0.25	0.25
Off-hook warning	1400	2450	0.1	0.1
Tone	2060	2600	0.1	0.1
Ring-back tone (US)	440	480	2.0	4.9
Ring-back tone (UK)	400	450	0.4	0.2
			0.4	2.0
Number unobtainable tone	400		Continuous	

2.4 Telephone Diallers

In modern telephone handsets, the old-generation rotary/pulse diallers are being replaced by touch-tone diallers. The disadvantages of pulse diallers are the following:

- The dialling time is dependent on the digits dialled; large digits (those close to 9, and 0) take maximum time.
- When the number of digits to be dialled is quite large (for example, in metros, the telephone numbers are 9- or 10-digits long), it takes enormous amount of time and effort.
- It is more prone to misdialling.

2.4.1 Touch-tone signals and DTMF/MFTD

Dual-tone multiple frequency (DTMF) has nearly completely replaced dial signalling since the advent of electrical switching offices. Touch-tone signalling uses a two-out-of-seven code to represent the 10 decimal digits and two other symbols (* and #).[4] There is another version of DTMF, that uses a 4 × 4 keypad matrix, with four additional keys marked 'A' for *Flash Override*, 'B' for *Flash*, 'C' for *Immediate*, and 'D' for *Priority*. These extensions to standard touch-tones originated with US military's Autovon Phone Network; they represented priority levels that could establish a phone connection with varying degree of immediacy, pre-empting other conversations on the network, if necessary, with Flash Override having the highest priority, down to Priority. Now these keys are mainly used in special applications such as amateur radio repeaters for their signalling or control. Individual frequencies in DTMF are so chosen that it is easy to design frequency filters and they can easily pass through telephone lines. DTMF standards specify 50 ms tone and 50 ms space duration. The keypad arrangement and the frequency allocation is illustrated in Figure 2.3. This arrangement is referred to as DTMF

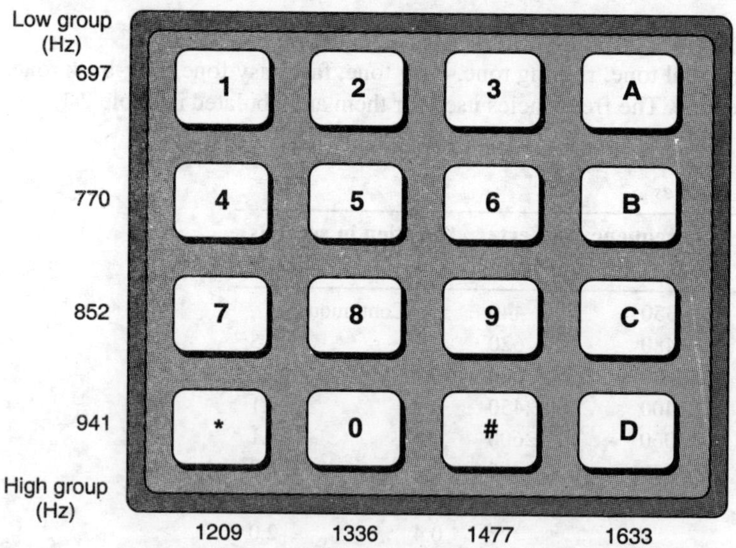

Figure 2.3 *Touch tone dialler (DTMF)*

[4]Source: *IT & T Reference Data for Radio Engineers*, 5th edition, Howard W. Sams & Co. Inc., 1969.

or multi frequency tone dialling (MFTD). Each of the eight states is designated by the presence of two separate tone frequencies within the voice band. The rotary telephone dialler unit is replaced by a keypad with 16 push-button switches. When one of the switches is depressed, two of the eight tones generators are energised and the two tones, one from the high group and one from the low group, are sent out audibly on the voice channel. The tone recognition equipment at the switching centre decodes the called number and sets up the selection switches.

Transmission of DTMF

Most often, a dedicated telephony circuit is used to generate DTMF. For example, the MT8880 or a microcontroller system is used. Just connecting an RC filter to the two output pins will generate the correct tones via a software. However, getting the exact frequencies often requires the use of a suitable crystal for the processor itself.

Decoding of DTMF

It is not very easy to detect and recognise DTMF with precision. Usually dedicated ICs are used, although a functional solution for DTMF transmission and reception by means of a microcontroller (a PIC in most cases) exists[5]. However, it is complicated. Most often, an MT8870 or a compatible circuit is used. MT8870 uses two sixth-order bandpass filters with switched capacitors. These produce a clean sine wave even from distorted inputs with harmonics suppressed. Most of the ICs do not require more than one 3.5 MHz crystal or resonator and a power circuitry. Usually, the output is four-bit binary plus a strobe.

2.5 Electronic Telephone

Electronic versions of the telephone provide the same functions as the traditional TS, but the heavy electromagnetic components are replaced by a few ICs. The circuit schematic of a typical electronic telephone is given in Figure 2.4. The circuit consists of just three ICs and few discrete components. Provision for both pulse and tone dialling are there.

2.6 Repeater Stations

Repeaters are amplifiers that are inserted in transmission lines at intervals to amplify the signal and compensate for transmission loss on the line. In the case of copper wire local loop, repeater spacing will be about 2 to 3 km. As the loss in optical fibres is less than 0.15 dB/km, in the case of optical fibre links repeaters the spacing can be as high as 60 to 90 km.

2.6.1 Two-wire transmission

In the case of two-wire transmission lines, transmission occurs in both directions, and so amplifiers used must be bilateral, amplifying equally well in both directions. *Negative impedance amplifiers* employing a principle similar to the negative resistance in oscillators is used. Negative impedance amplifiers are two-terminal bilateral devices. Before these became available, only unilateral amplifiers could be used and it was necessary to split the two-way transmission into two separate paths (four-wire) before the amplifiers could be inserted. Hybrid transformers or their electronic equivalents are required to divide the two-wire path into a four-wire path and return it to two-wire. Figure 2.5 shows a typical arrangement of a two-wire repeater.

[5] The acronym PIC stands for 'programmable, intelligent computer'.

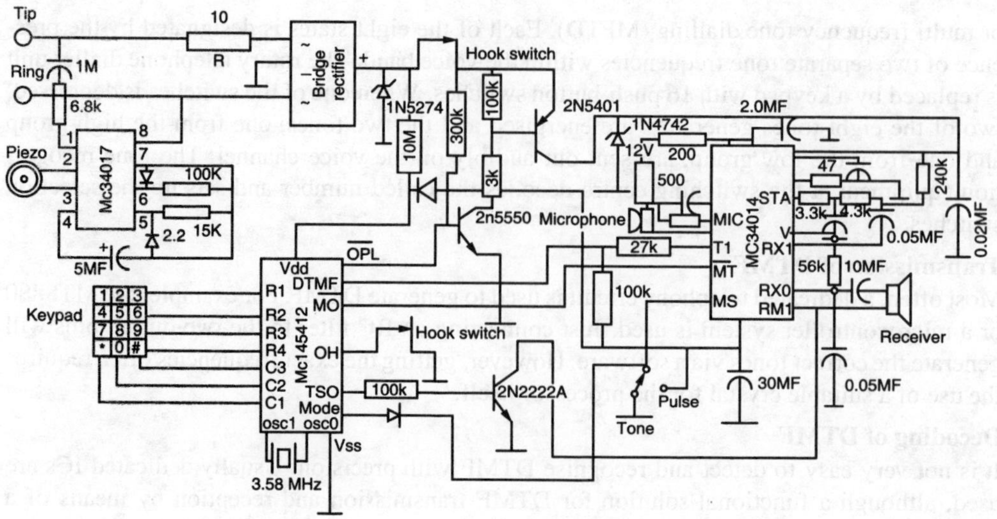

Figure 2.4 *Electronic telephone circuit*

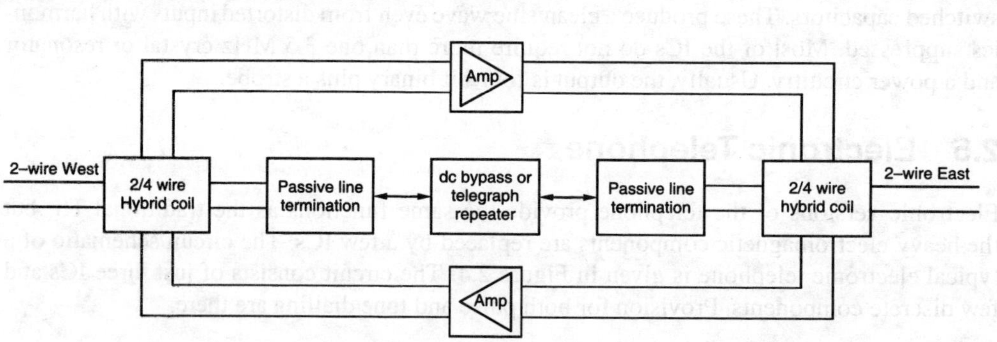

Figure 2.5 *Two-wire repeater circuit*

2.6.2 Four-wire transmission

Two-wire circuits are difficult to balance accurately, and the maximum obtainable gain is limited. For this reason, long-distance and interoffice trunk circuits are always done on a four-wire basis, with two wires providing transmission in one direction and the other two in the reverse direction. When carrier circuits are used, four-wire (two-path) transmission must be used because the carrier amplifiers are unilateral and allow only one-way transmission. Figure 2.6 shows a typical arrangement of equipment along the path of a four-wire long-distance circuit. The region marked by the dotted line indicates the boundary of the four-wire circuit, which extend over several kilometres, often covering different geographical regions. Note that the *local loop* is primarily a two-wire circuit. The customer's loop at each end is switched on a two-wire basis into a toll (long-distance) switching centre. A four-wire terminating set at the toll centre converts the circuit to the two-path mode. The long-distance facilities between the toll centre A and the far-end toll centre B are all four-wire and may include cable circuits, repeaters, carrier circuits, and radio circuits (including satellite links). At the far-end toll centre

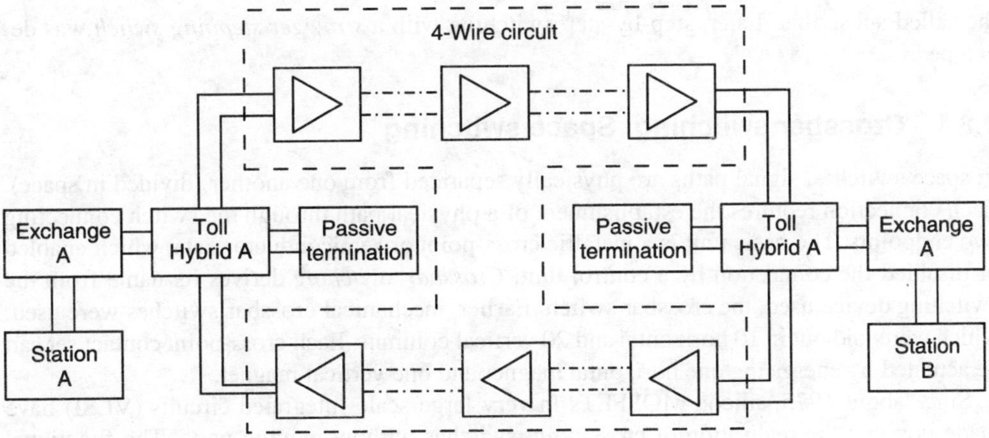

Figure 2.6 *Four-wire trunk circuit*

the transmission is converted back to two-wire before connecting to the far-end customer's loop.

2.7 Subscriber Loop Lines

The connection to subscribers within a central office is limited by the dc resistance of the loop pair. If the loop resistance is greater than 1200 Ω, the dc signalling becomes unreliable. If one uses 26 AWG copper, the aforementioned value of resistance limits the distance to about 5 km, and to 45 km if one uses 16 AWG copper.

A further limitation is placed on the loop length by the attenuation factor of the cable for voice signals. Loops with losses more than 10 dB total are not acceptable, and repeater amplifiers must be placed in the loop to make-up for the losses. As uncompensated or unloaded cable exhibits an increasing attenuation with increasing frequencies, it is therefore necessary to compensate for these in longer loops. The correction is known as *loading* and consists of placing series inductances at intervals along the cable. These inductances act with the cable capacitances to create a distributed lowpass filter with a cut-off frequency of about 3.5 kHz and a flat attenuation function over the frequency range of 300 Hz to 3300 Hz. The attenuation factor tends to be lower than that of uncompensated cable. In M88 loading, 88 mH loading coils are placed at 2743 m intervals.

2.8 Switching Equipment

The heart of the telephone system is the switching fabric. In the first-generation telephones, switching was done manually by an operator at the central office. Automatic exchanges or dial services came into existence in the early 1900s. Their purpose was to eliminate the need for human telephone operators. Before exchanges became automated, operators had to complete the connections required for a telephone call. This arrangement is even now used in some small private telephone systems to supplement automatic switching. The next-generation telephone systems used mechanical switches to connect and disconnect the calling subscriber to the called subscriber. Later, step-by-step switching with a *strowger stepping switch* was developed.

2.8.1 Crossbar switching: Space switching

In space switches, signal paths are physically separated from one another (divided in space). Each connection requires the establishment of a physical path through the switch connecting two endpoints. The basic unit is a metallic cross-point or semiconductor gate, which enabled or disabled the connection by a control unit. *Crossbar switching* derives its name from the switching device used, the crossbar switch. Earlier, mechanical crossbar switches were used, which were laid out in 10 horizontal and 20 vertical columns. Each cross-point contact set can be actuated by energising one horizontal magnet and one vertical magnet.

Since about 1975, silicon MOSFETs in very large scale integrated circuits (VLSI) have made possible the realisation of cross-point switches without moving parts. The functional block diagram of such a 4 × 4 cross-point switch, viz. Motorola MC142100, is given in Figure 2.7.

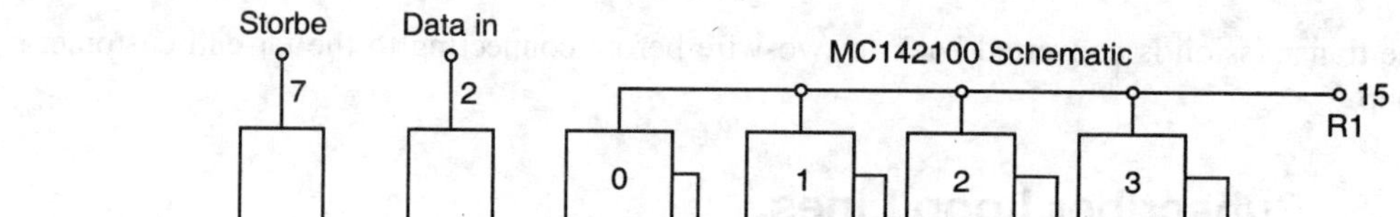

Figure 2.7 *A CMOS* 4 × 4 *crossbar switch*

A cross-point switch provides 16 analog transmission gates of the type shown in Figure 2.8. The switch can connect any of the 4 inputs to any one of the 4 outputs, and hence is called a 4 × 4 cross-point switch.

The maximum number of cross-points integrated into a chip is limited to about 16 × 16 because of the number of input and output pins required on the chip. A PCB containing 256 of such chips would form a completely non-blocking cross-point matrix for 256 × 256 circuits.

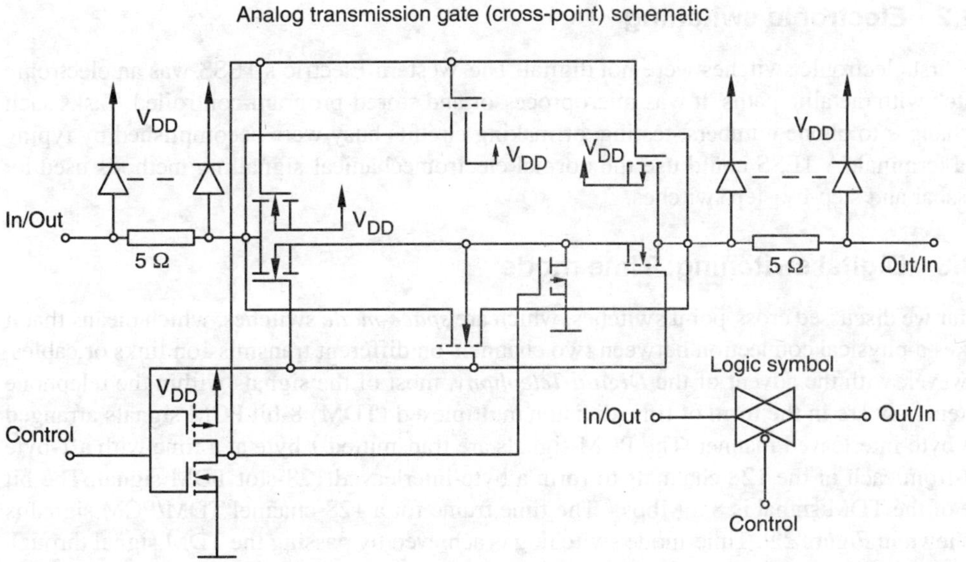

Figure 2.8 *Schematic of an analog transmission gate*

Cross-point switches are used in matrices to form space-mode switch units. If the matrix connects N input lines to N output lines, the matrix is square and is completely non-blocking. That is, N independent paths interconnecting the N inputs to the N outputs in any order can be established. With the square matrix, there are two cross-points with every path, resulting in more switches than are really necessary.

If the matrix is not square, i.e., say it has M inputs and N outputs, where $N < M$, then only N interconnections can be made at a time. Then the switching matrix need not be *strict-sense non-blocking.*[6] The value of N is fixed based on the statistical maximum demand for connection paths, or the *traffic* that can occur on the M inputs. More elaborate switch units incorporating two or more stages of cross-point switching are set up with enough cross-points to guarantee N simultaneous paths between M inputs and N outputs, where N represents the traffic level expected. If the demand for connections is close to N, then the probability of encountering a blocking condition rises significantly. The limitations of crossbar switches are enumerated below.

1. Number of cross-points increase as the square of number of stations.
2. Loss of cross-point prevents connection.
3. Inefficient utilisation of cross-points even when all the stations are active, i.e., only a small percentage of the cross-points are engaged.

One possible solution to the above problems is to use multiple-stage switching that reduce the number of cross-points required and increases the number of paths between two endpoints, which results in increased reliability. However, more control effort is required and internal blocking occurs sometimes (inability to connect two stations because all the possible paths are already in use).

[6]If a network can never have blocking, no matter what connections are present and with no need to arrange connections, the network is said to be *non-blocking in the strict sense*.

2.8.2 Electronic switching

The first electronic switches were not digital. The Western Electric's 1ESS was an electronic switch with metallic paths. It was microprocessor and stored-program-controlled. Tasks such as changes to phone numbers, testing, or making circuits busy were accomplished by typing on a terminal. A 1ESS could use the normal electromechanical signalling methods used by crossbar and step-by-step switches.

2.8.3 Digital switching: Time mode

So far we discussed cross-point switches, which are *space-mode* switches, which means that it makes a physical connection between two channels on different transmission links or cables. However, with the advent of the *Digital Telephony*, most of the signals within the telephone system now are in the form of time division multiplexed (TDM) 8-bit PCM signals arranged in a byte-interleaved manner. The PCM signals are transmitted 1 byte at a time with a 1-byte slot from each of the 128 channels to form a byte-interleaved 128-slot TDM signal. The bit rate of the TDM signal is 8.2 Mbps.[7] The time frame for a 128-channel TDM/PCM signal is as shown in Figure 2.9. Time-mode switching is achieved by passing the TDM signal through time slot interchanger (TSI), which interchanges the slot assignments within the signal. This necessitates receiving and temporarily storing the data bytes in a memory (shift-register) and then retransmitting them after a one-frame delay period (125 μs), but in a different order altogether. As an example, suppose a 1-byte data word is received from a line card in slot 55 during one frame. Slot 55 on the output side is already assigned (i.e. *busy*), and so a free slot must be found. The data byte is retransmitted in slot 143 on the outgoing line, which is assigned to a free input on the next switching stage.

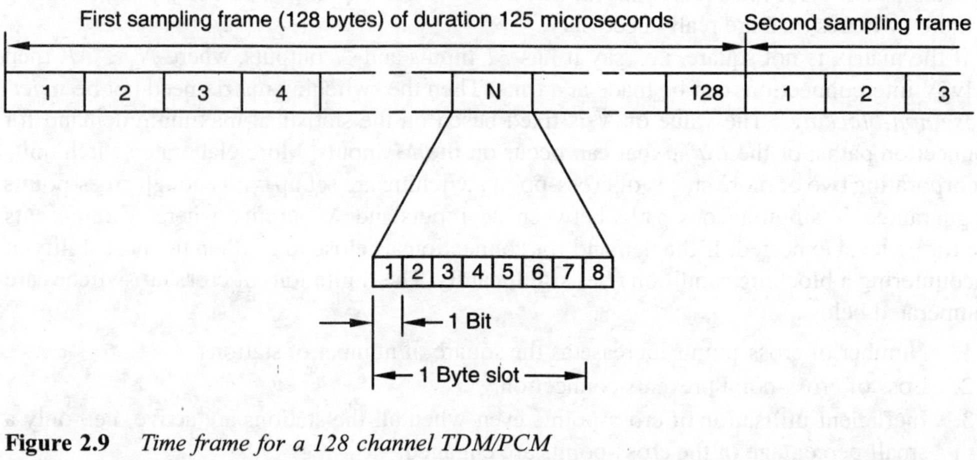

Figure 2.9 *Time frame for a 128 channel TDM/PCM*

The memory must be 8 bits wide (byte-oriented) and provide 2 bytes for each two-way transmission channel connected to it. It must also be sufficiently fast to accommodate the bit rate.

[7] As the analog telephone quality speech has a bandwidth of roughly 3.3 kHz, the sampling must be done at $> 2 \times 3.3$ kHz (8 kHz is chosen). As an 8-bit ADC is sufficient and 128 such speech channels are interleaved in a frame, the bit rate would be 8×8 kHz $\times 128 = 8.196$ Mbits/second.

A time switch does not necessarily have to have the same total number of slots available at both the input and the output. If it has N input slots and N output slots and N words of memory, then the time switch is completely *non-blocking*. This means that N separate paths can interconnect the N inputs and the N outputs in any order. If, however, it has N inputs and M outputs, such that $M > N$ and with N memory words, then only N paths can be connected at any time and one of the M inputs may be blocked; that is, it will encounter a busy condition. On the other hand, if the switch has M inputs and M outputs but only N memory words ($N < M$), again, only N paths can be established and the switch can present a blocking condition. The number of output slots and memory words provided in a given unit is based on the statistical probability of the maximum demand for path connections or the traffic that the switch must handle.

Digital switches work by connecting two or more digital virtual circuits together, according to a dialled telephone number. Calls are set up between switches using the Signalling System 7 protocol or one of its variants. In the US and in military telecommunication, a digital switch is a switch that performs time-division multiplexing switching of digitised signals.

2.8.4 Time–space switching

If a time switch is used independently, it needs a time-division multiplex at its input and a demultiplexer at its output. However, if a time switch and space switch are used together, then the space switch may also be time shared, resulting in a significant reduction in complexity of the switching fabric as regards the amount of hardware. Hence, most modern digital switching machines use a compound configuration that involves two time switches, with one or more space switches forming an interconnection matrix between them. As an example, one can cite the No.4 ESS electronic switching machine developed by Western Electric for AT&T Bell Labs. It has a T–S–S–S–S–T configuration which allows the interconnection of 100,000 telephone channels. A typical AT&T Bell System's 5ESS digital switch is shown in Figure 2.10.

2.9 Trunk Circuits

Trunk circuits interconnect two different switching centres. These switching centres may be two adjacent exchanges, or they may be separated by a considerable distance; a connection will involve a chain of several *toll (long-distance)* circuits over different facilities (radio, cable, and so on). Exchange interoffice circuits are mere extensions of loop circuits connected at either end. They must provide a two-way voice channel, which can be either a two-way (two-wire) circuit or two one-way circuits, one in each direction (four-wire), and they must provide a two-way channel for supervisory control signals such as the dial information and busy signals. Although two-wire circuits operating in the same way as loops may be used, four-wire circuits are preferred for interoffice connections. Carrier systems are more commonly used (discussed in section 2.11) so that fewer cable pairs between offices are sufficient. The objective is that the total transmission line loss between the points where the two loops are connected should approach 0 dB. Each loop is allowed a maximum loss of 3 to 5 dB so that the net connection loss measured at 1000 Hz is no more than 6 to 10 dB.

Toll connecting and inter-toll trunks are always designed on a four-wire basis. Ideally, the total transmission loss from end to end of any toll connection should be 0 dB. But as the two ends are terminated on a two-wire system which does reflect energy, it is possible to get an

Figure 2.10 *The AT&T 5ESS digital switch*

echo returning from the far end. This can result either in oscillations (if the loss is too low) or an annoying echo if the delay is more than about 20 ms.

Toll trunk circuits are designed on the *via net loss* (VNL) principle so that each circuit has a minimum amount of return loss built into it to keep the echo within acceptable limits. The VNL required for a toll circuit is given by:

$$\mathrm{VNL} = \frac{0.2 \times L}{V} + 0.4\,\mathrm{dB}, \tag{2.9.1}$$

where V is the velocity of propagation in km/ms, L the circuit length in kilometres and 0.2 the dB loss that must be inserted per millisecond of delay to obtain acceptable echo levels.

If the total VNL for a circuit is more than 2.5 dB (corresponding to 20 ms of round-trip delay time), devices called *echo suppressors* are to be inserted in the channel. They are voice-operated relays that sense the high level of a near-end talker on the outgoing channel and insert a high loss in the return channel for the duration of the speech. They disconnect about 100 ms after speech ceases so that the far-end party can start talking. A second circuit on the return channel does the same for the transmission in the other direction.

The noise in trunk lines must also be kept under acceptable levels. Noise on long circuits must be kept below 44 dBrn (dB-reference noise level), and those on short- and medium-length circuits below 38 dBrn.[8]

[8]0 dBrn corresponds to −90 dBm (referenced to 1 mW of 1000 Hz pure tone), or to −88 dBm of white noise in the bandpass from 0 to 3000 Hz.

2.10 Private Telephone Networks

Private telephone networks are communication systems that are owned, leased, or operated by the companies that use these systems. They primarily allow the interconnection of multiple telephones within the private network with each other and provide for the sharing of telephone lines from a public telephone network. They can be anything from two or three handsets on a single loop to an international network involving several centres, each with hundreds of stations. Private networks can be operated on a fully-manual basis where all control and switching is performed by an operator; this exchange is called a private branch exchange (PBX). They can also be fully automatic, in which case they are called private automatic branch exchange (PABX). The PABX is usually set up so that calls between stations within the system can be established on an automatic switching basis by equipment similar to that used in the exchange offices of the PSTN. One or more dial trunks connect the switching system to a nearby public exchange and it is possible for automatic dial-out (usually dialling '0' to get the external trunk). It is possible to have multiple trunk lines connected to the corporate PABX with the same number. Virtual private networks (VPN) are an extension of the concept of a PABX and are used by large companies.

2.11 Powerline Carrier Telephony

By now, it is known to the reader that with baseband audio transmission, only one pair of speech can be transmitted over a pair of conductors. On the other hand, if the baseband is used to modulate a high-frequency carrier, we can multiplex several audio bands into a single physical channel. This technique is referred to as frequency-division multiplexing (FDM). In powerline carrier telephony, the same powerline conductors that carry the 50 Hz ac power signal are also used to carry one or more radio frequency (RF) carrier modulated speech signals. They are usually used as a means of communication between the ac power stations and the transmission and distribution nodes. As the audio spectrum is elevated to high frequencies by modulation, they can be easily decoupled at the receiver using a '*wave trap*'.

2.12 Facsimile Transmission

Transmission of an exact replica of a hard-copy document, including legal signatures, is one of the most important features of a modern office. This is especially true for legal contracts, medical records, authorisations, etc. Facsimile ('fax' for short) machines have evolved through four generations (Groups 1, 2, 3 and 4). Groups 1 and 2 were analog machines; group 3 is the new analog machine. A comparative performance chart is given in Table 2.2. Group 4 machines transmit digitally. The older analog machine can take 6 minutes to transmit one 21.59×27.94 cm page. Group 3 machine can transmit the same page in less than a minute (20 s). The total time to send a page including the call setup time is 1 minute for group 3 fax. The top speed is 9600 bps for group 3 and 64 kbps for group 4.

Once a facsimile device is connected to a communication circuit, the following basic steps take place during transmission:

- Establish the call either manually dialling your telephone or by having an automatic call placed (physical circuit connection).
- The fax machine handles the pre-message procedure or *handshaking*, which consists of identifying the called station (facsimile machine) and any other procedures that might

Table 2.2 *The comparative performance of facsimile machines*

Type of machine	Speed (bps)	Resolution	Data compression
Group 1 and 2 Analog	300–1200	203 × 98 dpi	No
Croup 3 Analog	300–9600	203 × 98/196/391 dpi	Yes
Group 4 Digital	300–64000	400 × 400 dpi	Yes

be required to set up the station.

- The fax machine transmits the message to the other end (the session), which involves synchronisation between the two devices, any error detection and correction methodologies, and movement of the message from one facsimile device to the other.
- The fax machine completes the post-message procedure, which includes END OF MESSAGE signalling at the conclusion of the page, any signalling that signifies more than one page, end of transmission, and anything else required to end the session (physical circuit disconnected).

With a Fax/Voice switch one can connect a telephone and a fax machine to the same telephone local loop and save on the cost of maintaining an additional line for one of them. When a call comes in, the switch automatically connects to the facsimile machine or rings the voice telephone, depending on which is appropriate. If the incoming call is to the fax, most group 3 machines transmit a 1.1 kHz tone every 3 s. This is called the CNG (CalliNG) tone, and when the Fax/Voice switch hears it, it (the switch) automatically routes the call to the fax. Otherwise no tone is sent and the telephone rings.

With the advent of inexpensive personal computers, a fax card or fax modem may be connected to it and the PC can be used to send and receive fax messages[9]. With the Internet becoming ubiquitous, one more advancement has occurred. Now it is possible, with very little expense, to send and receive fax messages over the Internet using TCP/IP.

2.13 Modems

In this section we will examine the functional details of *modems*. A modem (a combinational word formed from modulator and demodulator) is a device that modulates an analog carrier signal to encode digital information; it also demodulates such a carrier signal to decode the transmitted information. The goal is to produce a signal that can be transmitted easily and decoded to reproduce the original digital data. Modems can be used in any means of transmitting analog signals, from driven diodes to radio.

The most familiar example of a modem turns the digital 1s and 0 s of a computer into sounds that can be transmitted over the telephone lines of the plain old telephone system (POTS), and once received on the other side, converts those sounds back into 1s and 0s. Modems are generally classified by the amount of data they can send in a given time, normally measured in bits per second, or 'bps'. Far more exotic modems are used by Internet users every day, notably cable modems and asymmetric digital subscribers loop (ADSL) modems. In telecommunications, 'radio modems' transmit repeating frames of data at very high data rates over microwave radio links. Some microwave modems transmit more than a hundred million bits per second. Optical modems transmit data over optical fibres. Most intercontinental data links

[9]The current price (year 2007) of a fax modem in Indian market is below Rs. 500.

now use optical modems, transmitting over undersea optical fibres. Optical modems routinely have data rates in excess of a billion (10^9) bits per second. Modems are widely classified into four categories:

- Telephone modem
- Fax modem
- Radio modem
- Broadband (cable) modem

Telephone modems are used to interconnect the telephone line to computer, whereas fax modems interconnect fax machine to the same telephone line. Cable modems find application in linking the TV cable network to the computer. Note that Internet is made available over the TV cable network by some of the Internet service providers (ISP). Cable modems are vital for providing *broadband Internet* service to households.

Null modem

Null modem is a communication method to connect two data terminal equipments (DTEs, computer, terminal, printer etc.) directly using a RS-232 serial cable. The original RS-232 standard only defined the connection of DTEs with data communication equipments (DCEs), i.e., modems. With a null-modem connection the transmitting and receiving lines are cross-linked. Depending on the purpose, sometimes also one or more handshake lines are cross-linked. Several wiring layouts are in use because the null-modem connection is not covered by a standard.

Null modems are commonly used for file transfer between computers or for remote operation. Under the Microsoft Windows operating system, a direct cable connection can be used over a null-modem connection. The later versions of MS-DOS were shipped with the Interlink program. Both pieces of software allow the mapping of a hard disk on one computer as a network drive on the other computer. No networking hardware (such as a network interface card or a modem) is required for this. The popularity and availability of faster information exchange systems such as Ethernet made null-modem cables less common. Figure 2.11 shows a typical null modem.

2.13.1 The Smartmodem

A major advancement in modem technology has been the Smartmodem, introduced in 1981 by Hayes Communications. The Smartmodem was an otherwise-standard 103 A 300 bps modem, but was attached to a small controller that let the computer send commands to it to operate the phone line. The command set included instructions for picking up and hanging up the phone, dialling numbers, and answering calls. The basic Hayes command set remains the basis for computer control of most modern modems. Prior to the Smartmodem, modems almost universally required a two-step process to activate a connection: first, the manual dialling of the remote number on a standard phone handset, and second, the plugging of the handset into an acoustic coupler. Since the modem could not dial the phone, the acoustic coupler remained common because one needed to dial the phone separately anyway. Hardware add-ons, known simply as diallers, were used in special circumstances and was generally operated by emulating the dialling of a handset. With Smartmodem a computer could dial the phone directly by sending the modem a command. This eliminated the need for an associated phone to dial with and, in turn, the need for an acoustic coupler. The smartmodem, instead, plugged directly into

Figure 2.11 *A typical null modem adapter*

the phone line. This greatly simplified the setup. Terminal programs that maintained lists of phone numbers and would send the dialling commands became common. The important modem standards and their respective speeds are given in Table 2.3. Figure 2.12 shows a typical Motorola data modem.

Table 2.3 *Major modem standards*

Modem Standard	Speed in bauds	Technology
103A	300 (300 bps)	FDX, FSK, analog
202B	1200	HDX, FSK, analog
212A	600 (1200 bps)	FDX, QPSK, analog
V.22	600	FDX, PSK, analog
V.26	1200 (2400 bps)	FDX, DQPSK, analog
V.32	2400 (4800/9600 bps)	FDX, Trellis Coded Modulation
V.32 bis	14.4 kbps	FDX, TCM
V.32 ter	19.2 kbps	FDX, TCM
V.34	28.8 kbps	FDX,PCM
V.34 bis	33.6 kbps	FDX,PCM
V.42	28.8 kbps	FDX, Error Correction
V.42 bis	34 kbps	FDX, Data Compression.
V.90	56 kbps	FDX, PCM, digital
V.92	56 kbps	FDX, PCM, digital

2.13.2 Softmodem

A winmodem or a softmodem is a stripped-down modem for MS Windows that replaces tasks traditionally handled in hardware with software. In this case the computer's built-in sound hardware is used to generate the tones normally handled by the analog portion of the mo-

Figure 2.12 *A 28.8 kbps serial-port modem from Motorola*

dem. A small piece of hardware is then used to connect the sound hardware to the phone line. Modern computers often include a very simple card slot, the communications and networking riser slot (CNR), to lower the cost of connecting it up. The CNR slot includes pins for sound, power, and basic signalling, instead of the more expensive PCI slot normally used. Winmodems are cheaper than traditional modems as they have fewer hardware components. One downside of a winmodem is that the software generating the modem tones is not that simple, and the performance of the computer as a whole often suffers when it is being used. For online gaming this can be a real concern. Another problem with winmodems is the lack of flexibility because of their strong tie to the underlying operating system. A given winmodem might not be supported by other operating systems (such as Linux), because their manufacturers may neither support other operating systems nor provide enough technical data to create an equivalent driver. A winmodem might not even work (or work well) with a later version of Microsoft Windows if its driver turns out to be incompatible with the later version of the operating system.

2.13.3 Radio modems

Because signals travelling through the air must be analog (digital signals cannot be transmitted without a guided transmission medium such as wire), modems are required for any digital transmission through the air. Today's data transmission needs are almost entirely digital, and as such, most modern wireless technology includes some form of modem to produce an analog signal suitable for radio transmission. Direct broadcast satellite, GPS, WiFi, and mobile phones all use wireless modems to communicate, as do most other wireless services today. The modern telecommunications and data networks also make extensive use of radio modems where inexpensive long-distance data links are required. Such systems are an important part of the PSTN, and are also in common use for high-speed computer network links to outlying areas where fibre is not economical. Even where a cable is installed, it is often possible to get better performance or make other parts of the system simpler by using radio frequencies and modulation techniques through a cable. The coaxial cable has a very large bandwidth; however, signal attenuation becomes a major problem at high data rates if a digital signal is used. By using a modem, a much larger amount of digital data can be transmitted through a single piece of wire. Digital cable television and cable Internet services use modems to provide the increasing bandwidth needs of modern households. A modem also allows for frequency-division multiple access to be used, making FDX digital communication with many users using a single wire possible.

Wireless modems come in a variety of types, bandwidths and speeds. Wireless modems are often referred to as transparent or smart. They transmit information that is modulated onto a carrier frequency to allow many simultaneous wireless communication links to work simultaneously on different frequencies. Transparent modems operate in a manner similar to their phone line modem cousins. Typically, they are half-duplex, meaning that they cannot send and receive data at the same time. Transparent modems are polled in a round-robin fashion to collect small amounts of data from scattered locations that do not have easy access to wired infrastructure. Transparent modems are most commonly used by utility companies for data collection.

Smartmodems come with a media access controller inside that prevents random data from colliding, and re-sends data that is not correctly received. Smart modems typically require more bandwidth than transparent modems and achieve higher data rates. The IEEE 802.11 standard defines a short-range modulation scheme that is used on a large scale throughout the world. Wireless data modems are used in the WiFi and WiMax standards, operating at microwave frequencies.

WiFi could be used in laptops for Internet connections (wireless access point and wireless application protocol/WAP).

2.13.4 Broadband modem

Asymmetric digital subscribers loop (ADSL) modems, a more recent development, are not limited to the telephone's 'voice band audio frequencies. ADSL is a form of digital subscribers line (DSL), a data communications technology that enables faster data transmission over copper telephone lines than a conventional modem can provide. Some ADSL modems use coded orthogonal frequency-division modulation. Cable modems use a range of frequencies originally intended to carry RF television channels. Multiple cable modems attached to a single cable can use the same frequency band, using a low-level media access protocol to allow them to work together within the same channel. Typically, 'up' and 'down' signals are kept separate using FDM. New types of broadband modems are beginning to appear, such as doubleway satellite and powerline modems.

Broadband modems should still be classed as modems because they do utilise analog/digital conversion. They are more advanced devices than traditional telephone modems as they are capable of modulating/demodulating hundreds of channels simultaneously. The photograph of a typical DSL modem is given in Figure 2.13.

2.14 Recent Trends

The recent trends in telephony include leased lines, independent telephone systems including central exchange (CENTREX), computer telephone integration (CTI), time-assigned speech interpolation (TASI), integrated-services digital network (ISDN), automatic call distribution (ACD) and interactive voice response system (IVRS). Enhancements also include IP LAN telephony and multimedia communications. More recent services aimed at the private network market are VPN and management network services (MNS). Voice-over IP (VoIP) telephony and facsimile-over IP (FoIP) are another two major developments.

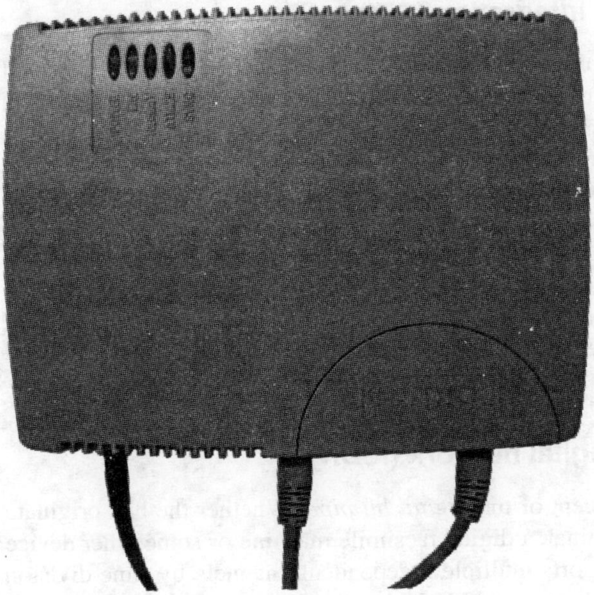

Figure 2.13 *ADSL modem from Motorola*

2.14.1 Leased lines

Leased lines (also known as private circuits/private wires or hot lines) give the user a permanent (or semi-permanent) connection between two endpoints, say from a branch office to the head office. No per-call switching is involved. Users pay a fixed monthly or annual charge irrespective of the usage. The annual charge will depend on the capacity and distance between the points connected. Leased lines were originally all analog, but now they are largely digital. The lines vary in their capacity, from a basic, permanently wired phone connection to a 155 Mbps permanent digital connection.

2.14.2 Centrex or Featureline

Centrex offers customers a 'virtual' PABX which routes their internal telephone calls through the local telephone exchange but charges a flat monthly fee per line. The customers do not have to buy and maintain a PABX, thereby saving space at their site. Upgradation is also not necessary since it will be taken care of by the telephone company. But Centrex services are becoming less popular with the advent of compact PABXs. *Featureline* is a proprietary Centrex product of British Telecom.

2.14.3 Computer–telephone integration (CTI)

This basically enables the use of telephone line for voice and data traffic using the personal computer in conjunction with the phone. Dialling can be done from the computer. It needs a 'set-top-box' to be installed on the computer.

2.14.4 Time-assigned speech interpolation (TASI)

TASI is used on some of today's long-distance voice lines. It allows for packing of extra voice conversations into a fixed number of circuits. Usually when two people engage themselves in a conversation, both parties do not speak at the very same moment, and for a small portion of the time neither speaks. This principle is effectively used to interpolate (interweave) several conversations on the same channel.

2.14.5 Integrated services digital network (ISDN)

The key idea behind ISDN is the concept of the *digital bit pipe*. Whether the bits originate from a digital telephone, a digital terminal, a digital facsimile machine or some other device is irrelevant. The digital bit pipe supports multiple independent channels by time-division multiplexing (TDM) of the bit stream. In short, in ISDN the signalling and voice channels are independent and digital.

2.14.6 Automatic call distribution (ACD)

Automatic call distribution (ACD) system uses an interactive voice response system (IVRS) to determine call routing. When an incoming call is initially received, the ACD system coordinates with the IVRS to determine the customer's selection. The ACD system then looks into the databases to retrieve the customers' account or other relevant information and transfers the call through the PBX to a qualified customer service representative (CSR). The ACD system may also transfer customer or related product information to the CSR.

2.14.7 Interactive voice response system (IVRS)

The IVRS is commonly used in public information systems such as the railroad enquiry or airline enquiry. When the customer dials the IVRS number, he/she is directed to different options available by an automatic voice response system. Often they have the option to choose the language as well.

2.14.8 IP LAN telephony

The internet protocol LAN telephony can be either IP-enabled or IP-centric. Under the IP-enabled method, IP telephony gateway cards can be added to existing PBXs, which will enable them to interface with IP phones and soft phones connected to local and remote LANs. This approach provides enterprises with the option of preserving much of their investment in digital and analog phones and the corresponding line and trunk circuit cards. The IP-enabled system at a central site can support users at a remote site without having to install communications servers and applications servers at the remote site. Media gateways are installed at the remote site to provide connectivity to the PSTN so that local calls do not have to be back-hauled over the IP WAN to the central site. Since all of the call processing is performed at the central site, the remote site would be unable to make calls if the IP WAN were disabled or congested. Consequently, the remote site requires a set of telephony features for survivability purposes.

Under the IP-centric method, the PBX is replaced with a communications server, which executes the call processing software. This method has been perceived as essentially a pure-IP approach, whereby all of the existing phones are replaced by IP phones or soft phones in which the PC serves as the phone. However, the IP-centric approach does utilise gateways with voice interface cards to support analog devices, such as fax machines and basic analog phones. Gateways are also used by IP-Centric systems to connect to the PSTN via either analog or digital trunks. Some IP-centric systems also provide digital gateways that enable these LAN-based systems to support proprietary digital phones.

2.14.9 Virtual private networks (VPNs) and MNS

These extend the Centrex concept to the entire public network, offering something that resembles a private network but actually carries the traffic more cheaply on the public network. The network operator also manages and maintains the VPN. A further benefit of MNS is that customers can ask for additional capacity as and when needed.

2.14.10 Voice-over IP telephony (VoIP)

Voice-over Internet Protocol (VoIP) technology results in huge savings on the amount of physical resources needed to communicate by voice over long distances. As data traffic continues to increase at a much faster pace than voice/telephone traffic, the interest in consolidating voice and data networks has been considerable. While previously the emphasis had been on transporting data over voice networks and using modems to convert digital data to analog phone signals, the current emphasis is on providing voice services over data networks. Data networks are optimised for carrying data and are important for phasing out the existing voice networks. As it makes use of the ubiquitous IP protocol, VoIP has become a leading technology for implementing voice services over data networks. Businesses are increasingly switching over to VoIP systems. Frost and Sullivan, a marketing research firm, recently estimated the annual growth rate for IP-enabled telephone equipment to have been 132% during the period 1997 to 2002. Another estimate says that 70% of Fortune 1000 companies use VoIP in their network structure. VoIP can provide a number of benefits to telecom/Internet service providers. Consolidating both data and voice networks provides potential cost savings to agencies. Communicating over data rather than voice system cuts down on long-distance charges between agencies. This infrastructure also allows smaller agencies with limited operating hours to roll over their services to larger call centres that have expanded hours. Additionally, VoIP can provide an infrastructure that allows for a statewide integration of resources and databases. Finally, a VoIP system provides transparency to callers.

2.14.11 Facsimile-over IP (FoIP)

FoIP is a fax-over Internet Protocol application that enables the interworking of standard fax machines with packet networks. It accomplishes this by extracting the fax image from an analog signal and carrying it as digital data over the packet network.[10] Traditionally, there have

[10]More information at http://www.iec.org/online/tutorials/vfoip/topic07.html

been two approaches for sending FoIP networks: real-time methods and store-and-forward methods. The primary difference in service between these two approaches lies in the delivery and method of receipt confirmation. The Frame Relay Forum has defined a real-time protocol for the transmission of fax-over frame relay networks. Likewise, the International Telecommunication Union (ITU) and Internet Engineering Task Force (IETF) worked together to evolve both the real-time FoIP network standard (T.38) as well as the store-and-forward FoIP network standard (T.37). Both T.37 and T.38 were approved by the ITU in June 1998.

There are tremendous opportunities for cost savings by transmitting fax calls over packet networks. Fax data in its original form is digital. However, it is modulated and converted to analog for transmission over the PSTN. This analog form uses 64 kbps of bandwidth in both directions.

2.14.12 Wireless in local loop telephony

In a basic telephone system, the access network between the local telephone exchange and the subscriber's premises forms the *local loop*. Typically, this local loop is copper based. In WLL (wireless in local loop) systems, the copper local loop is replaced with a RF loop, maintaining the existing services of a basic telephone system.

Accessing WLL system

One of the popular ways for accessing the WLL system by users is code-division multiple access (CDMA). In CDMA, all users communicate simultaneously through the same radio channel, occupying the entire channel bandwidth. The distinction between users is made on the basis of a unique code assigned to each user. Other methods of accessing WLL are FDMA and TDMA. The frequency bands allotted for CDMA-based WLL systems is 824–849 MHz (uplink) and 869–894 MHz (downlink).

Cell concept

WLL being a radio communication system, divides the entire coverage area into a number of hexagonal sub areas called *cells*. A cell may be either micro or macro in nature. A microcell has a radius of 50 to 200 m, while that of a macrocell is about 10 to 30 km. The CDMA-based WLL system uses macrocell architecture.

WLL users have limited mobility compared with mobile phone users. A WLL customer usually gets services within his/her cell area, while mobile phone users can move from one cell to another.

System architecture of CDMA-based WLL

The block diagram of a CDMA-based WLL system is shown in Figure 2.14. The basic subunits of a WLL system are base station controller (BSC), base transreceiver subsystem (BTS), operation and maintenance centre (OMC), and mobile set and fixed wireless terminal. The BSC, which is the heart of the WLL system, interfaces with the PSTN, BTS, and OMC to perform call processing, voice coding, and operation- and maintenance-related tasks. The BTS is the interface between the BSC and subscriber stations. It performs tasks related to wireless resource management, power control, call processing and testing. The BTS has an antenna

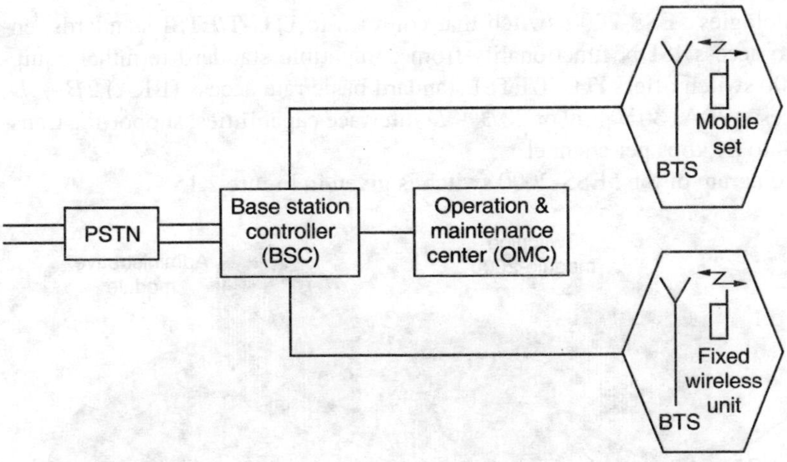

Figure 2.14 *CDMA-based WLL system architecture*

system which may be omnidirectional or directional. Normally, BTS lies at the centre of the hexagonal cell and serves subscribers within the cell.

Subscribers may be provided with either a portable telephone or a fixed wireless terminal. The portable mobile telephone is basically a mobile set. The fixed wireless terminal consists of a telephone instrument and a radio terminal. There is provision to connect a fax machine or a PC to the fixed wireless terminal in addition to the telephone instrument. The mobile set works on a 3 V rechargeable, while the fixed wireless terminal has a 230 V/12 V adapter in addition to a 12 V rechargeable maintenance-free battery.

Advantages of WLL system

The following are the advantages of WLL system.

1. The cost of establishment of new connections is drastically reduced in unfriendly terrains and the time required to route new links is also very less.
2. In a fixed telephony environment, CDMA WLL provides upto 15–20 times the capacity of advanced mobile phone cellular system (AMPS).
3. It requires less power and efficient power control is possible.
4. CDMA optimises use of the radio spectrum.
5. It results in improved *quality of service* (QoS) and call quality.

2.15 ISDN Exchanges

The integrated-services digital network (ISDN) is a set of international recommendations and standards promoted by the International Telecommunications Union–Technical Sector (ITU-T) and the European Telecommunications Standards Institute (ETSI) to standardise telecommunications signalling and transfer protocols. ISDN enables simultaneous high speed voice, text, graphics, and video transmission over ordinary telephone lines. Offering end-to-end digital connectivity for 64 kbps digital channels, ISDN provides users with access to both circuit-switched and packet-switched network applications. There are several implementations such as the Lucent Technologies 5ESS-2000 switch that conform to ITU-T/ETSI standards, enabling subscribers to access ISDN functionality from compatible standard terminal equipment. The 5ESS-2000 switch offers ITU-T/ETSI standard basic rate access (BRA) $2B + D$, and primary rate access (PRA) $30B + D$ or $23B + D$ interface capabilities, supporting transmission speeds of up to 64 kbps per channel.

The connectivity diagram of the 5ESS-2000 switch is given in Figure 2.15.

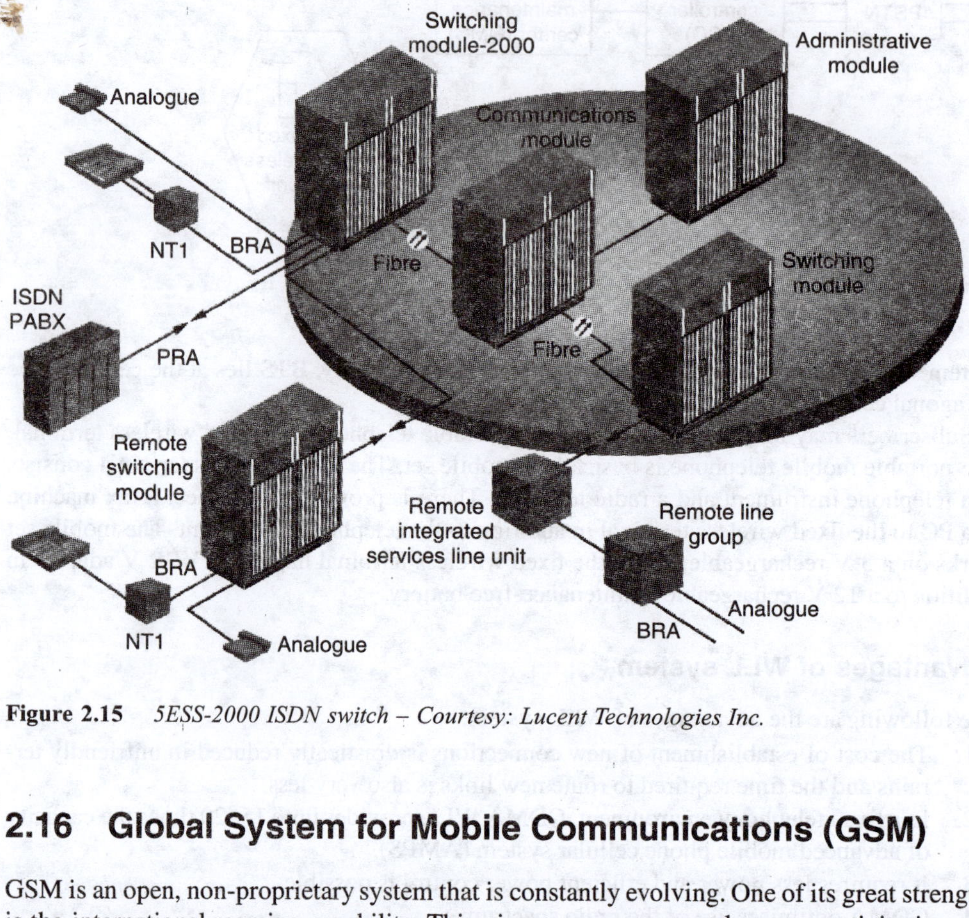

Figure 2.15 *5ESS-2000 ISDN switch – Courtesy: Lucent Technologies Inc.*

2.16 Global System for Mobile Communications (GSM)

GSM is an open, non-proprietary system that is constantly evolving. One of its great strengths is the international roaming capability. This gives consumers seamless connectivity in more than 170 countries over a single subscriber number. GSM satellite roaming has extended service access to areas where terrestrial coverage is not available.

GSM differs from first-generation wireless systems in that it uses digital technology and TDMA transmission methods. Voice is digitally encoded via a unique encoder, which emulates the characteristics of human speech. This method of transmission permits a very good data rate to information content ratio.

High-bandwidth services are already becoming available through second-generation technologies. The development path to 3GSM (which is a generic term for the next-generation mobile technology) is clearly mapped out and brings with it the possibilities of sophisticated data and multimedia applications. The GSM standard will continue to evolve along with wireless, satellite, and cordless systems, offering greatly expanded services. These will include high-speed, multimedia data services, in-built support for parallel use of such services and seamless integration with the Internet and wireline networks.

Video on demand, high-speed multimedia, and Internet access are just a few possibilities of the advanced services that 3GSM can offer. The main benefit of third generation systems is that they will offer high-end service capabilities which include substantially enhanced capacity, quality, and data rates greater than what is currently available. 3GSM services will also include concurrent usage of multiple services and they bridge the gap between wireless and Internet/computing.

The GSM Association, based in Dublin, Ireland, and London, UK, represents the interests of more than 690 GSM, satellite and 3GSM operators, key manufacturers, and suppliers to the GSM industry as well as regulatory and administrative bodies from more than 190 countries and regions around the world. Most of the first third generation licensees are also members. The GSM Association is responsible for the continued maintenance of open standards and interoperability. The global cooperation between operators is most powerfully illuminated by the success of international roaming. One of the association's major priorities is the development and promotion of the GSM standard worldwide.

Although the CDMA technology offer some competition to GSM technology, it will continue to be in service for many years to come.[11]

2.16.1 GSM technology

GSM is a cellular network, which means that mobile phones connect to it by searching for cells in the immediate vicinity. GSM networks operate at different radio frequencies. Most GSM networks operate in the 900 MHz or 1800 MHz bands. Some networks in parts of the Americas (including the US and Canada) operate in the 850 MHz or 1900 MHz bands because the 900 MHz and 1800 MHz frequency bands were already allocated to other users. An even smaller number of areas use the 400 MHz and 450 MHz frequency bands. Because of this proliferation of bands, there is no single phone sold that can work at full capabilities on all GSM systems in the world.

In the 900 MHz band the uplink frequency band is 890–915 MHz, and the downlink frequency band is 935–960 MHz. This 25 MHz bandwidth is subdivided into 124 carrier frequency channels, each spaced 200 kHz apart. Time-division multiplexing is used to allow eight speech channels per RF channel. There are eight radio time-slots (giving eight burst periods) grouped into what is called a TDMA frame. The channel data rate is 270.833 kbps, and the frame duration is 4.615 ms.

The transmission power in the handset is limited to a maximum of 2 watts in GSM850/900 and 1 watt in GSM1800/1900.

GSM uses linear predictive coding (LPC). The purpose of LPC is to reduce the bit rate. The LPC provides parameters for a filter that mimics the vocal tract. The signal passes through this filter, leaving behind a residual signal. Speech is encoded at 13 kbps.

There are four different cell sizes in a GSM network – macro, micro, pico, and umbrella cells. The coverage area of each cell varies according to the implementation environment. Macrocells can be regarded as cells where the base station antenna is installed on a mast or a building above average rooftop level. Microcells are cells whose antenna height is under average rooftop level; they are typically used in urban areas. Picocells are small cells whose

---

[11] http:// *www.gsmworld.com/technology/faq.shtml*
http:// *en.wikipedia.org/wiki/GSM*

diameter is a few dozen meters; they are mainly used indoors. On the other hand, umbrella cells are used to cover shadowed regions of smaller cells and fill in gaps in coverage between those cells.

Cell radius varies depending on antenna height, antenna gain, and propagation conditions from a couple of hundred meters to several tens of kilometres. The longest distance the GSM specification supports in practical use is 35 km or 22 miles. There are also several implementations of the concept of an extended cell, where the cell radius could be double or even more, depending on the antenna system, the type of terrain, and the timing advance. Indoor coverage is also supported by GSM and may be achieved by using an indoor picocell base station, or an indoor repeater with distributed indoor antennas fed through power splitters, to deliver the radio signals from an antenna outdoors to the separate indoor distributed antenna system.

The modulation used in GSM is Gaussian minimum shift keying (GMSK), a kind of continuous-phase frequency shift keying. In GMSK, the signal to be modulated onto the carrier is first smoothed with a Gaussian lowpass filter prior to being fed to a frequency modulator, which greatly reduces the interference to neighbouring channels (adjacent channel interference).

2.16.2 GSM architecture

A GSM network is composed of several functional entities whose functions and interfaces are specified. Figure 2.16 shows the layout of a generic GSM network. The GSM network can be divided into three major parts. The mobile station is carried by the subscriber. The base station subsystem controls the radio link with the mobile station. The network subsystem, the main part of which is the mobile services switching centre (MSC), performs the switching of calls between the mobile users and those between mobile and fixed network users. The MSC also handles the mobility management operations. Not shown is the OMC, which oversees the proper operation and setup of the network. The mobile station and the base station subsystem communicate across the *Um interface*, also known as the air interface or radio link. The base station subsystem communicates with the mobile services switching centre across the *A interface*.

Mobile station

The mobile station (MS) consists of the mobile equipment (the terminal) and a smart card called the subscriber identity module (SIM). The SIM card provides personal mobility so that the user can have access to subscribed services irrespective of a specific terminal. By inserting the SIM card into another GSM terminal, the user is able to receive calls at that terminal, make calls from that terminal, and receive other subscribed services.

The mobile equipment is uniquely identified by the international mobile equipment identity (IMEI). The SIM card contains the international mobile subscriber identity (IMSI) used to identify the subscriber to the system, a secret key for authentication, and other information. The IMEI and the IMSI are independent, thereby allowing personal mobility. The SIM card may be protected against unauthorised use by a password or personal identity number (PIN).

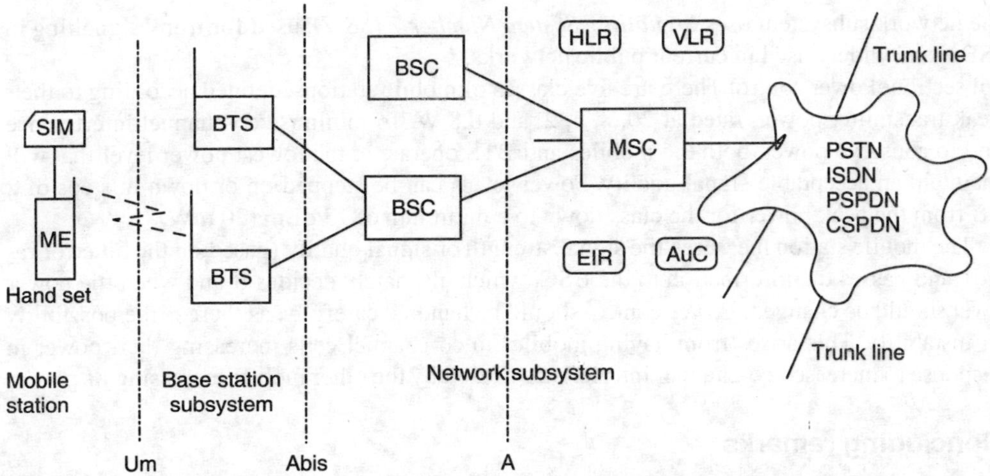

PSTN : Public Switched Telephone Nnetwork
ISDN : Integrated Services Digital Network

SIM : Subscriber Identity Module
ME : Mobile Equipment
BTS : Base Transceiver Station
BSC : Base Station Controller
HLR : Home Location Registrer

PSPDN : Packet Switched Public Ddata Network
CSPDN : Circuit Switched Public Data Network

VLR : Visitor Location Register
MSC : Mobile Services Switching Centre
EIR : Equipment Identity Register
AuC : Authentication Centre

Figure 2.16 *The GSM architecture*

Base station subsystem

The base station subsystem is composed of two parts, the base transceiver station (BTS) and the base station controller (BSC). These communicate across the standardised *Abis interface*, allowing (as in the rest of the system) operation between components made by different suppliers.

The BTS houses the radio tranceivers that define a cell and handles the radio-link protocols with the mobile station. In a large urban area, there will potentially be a large number of BTSs deployed, and thus the requirements for a BTS are ruggedness, reliability, portability, and minimum cost.

The BSC manages the radio resources for one or more BTSs. It handles radio-channel setup, frequency hopping, and handovers. The BSC is the connection between the mobile station and the mobile service switching Centre (MSC).

Network subsystem

The central component of the network subsystem is the MSC. It acts like a normal switching node of the PSTN or ISDN, and additionally provides all the functionality needed to handle a mobile subscriber such as registration, authentication, location updating, handovers and call routing to a roaming subscriber. These services are provided in conjunction with several functional entities, which together form the network subsystem. The MSC provides the connection to the fixed networks (such as the PSTN or ISDN). Signalling between functional entities in the network subsystem uses *Signalling System Number 7 (SS-7)*, used for trunk signalling in ISDN and widely used in current public networks.

subsection*Power control There are five classes of mobile stations, defined according to their peak transmitter power, rated at 20, 8, 5, 2, and 0.8 W. To minimise co-channel interference and to conserve power, both the mobiles and BTS operate at the lowest power level that will maintain an acceptable signal quality. Power levels can be stepped up or down in steps of 2 dB from the peak power for the class down to a minimum of 13 dBm (20 mW).

The mobile station measures the signal strength or signal quality (based on the bit error ratio), and passes the information to the BSC, which ultimately decides if and when the power level should be changed. Power control should be handled carefully, as there is the possibility of instability. This arises from having mobiles in co-channel cells increasing their power in response to increased co-channel interference caused by the other mobile increasing its power.

Concluding remarks

GSM has shown its commitment to openness, standards and interoperability, and is compatible with ISDN which is evolving in most industrialised countries. GSM is also the first system to make extensive use of the intelligent networking concept, in which services such as 800 numbers are concentrated and handled from a few centralised service centres, instead of being distributed over every switch in the country. This is the concept behind the use of the various registers such as the HLR. In addition, the signalling between these functional entities uses SS-7, an international standard already deployed in many countries and specified as the backbone signalling network for ISDN.

GSM is a very complex standard, but that is probably the price that must be paid to achieve the level of integrated service and quality offered under the rather severe restrictions imposed by the radio environment.

Appendix: Circuit Switching

Telephone traffic and erlangs

Trunks and links, being busy, constitute independent random events. If two random events are independent, the probability of both happening at the same time is given by the product of their separate probabilities of occurrence at that time. This assumption is called the independence assumption. For example, if two links are to be connected in tandem and the probability of one being busy is a and of the other being busy is b, then the probabilities of each being free are $1 - a$ and $1 - b$, respectively; so the probability of both being free is $(1 - a)(1 - b)$. Therefore, the probability of the path being blocked is $1 - (1 - a)(1 - b)$.

The occupancy at each stage is the total traffic carried divided by the number of links at that stage. However, if the loss is small (as it should be), a little error is introduced by using the traffic offered instead of the traffic carried. On a group of trunks, the average number of calls in progress depends on both the number of calls that arrive and their duration. The duration of a call is often called its holding time because it holds a trunk for that time. The traffic in erlangs is equal to the mean number of calls arriving during a period equal to the mean duration of the calls.

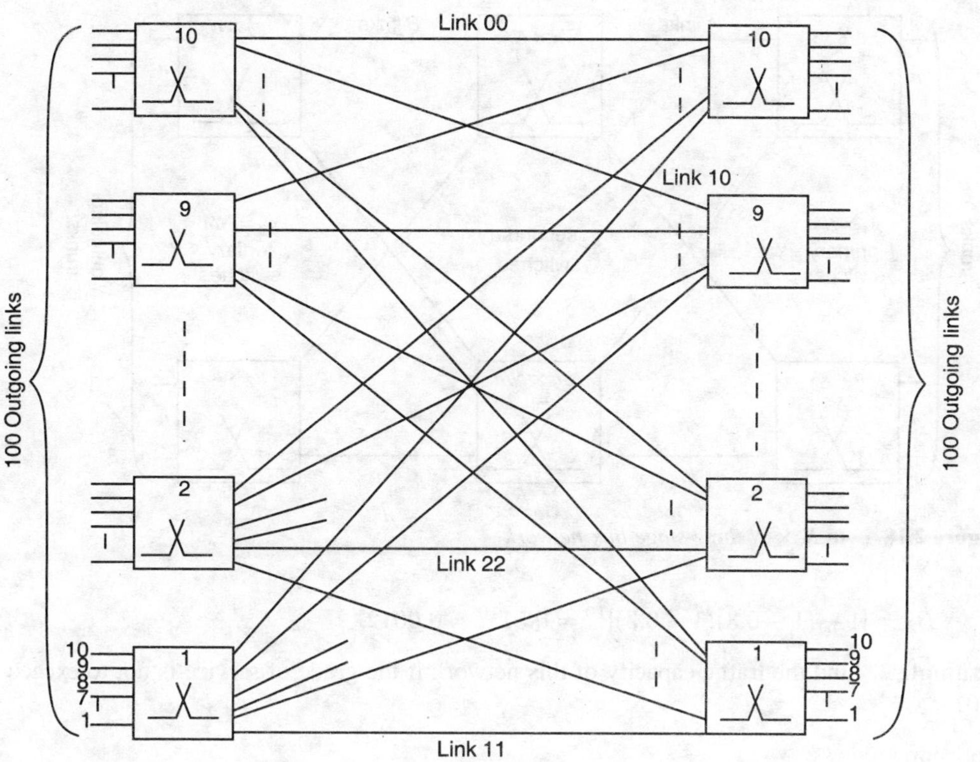

Figure 2.17 *A 100 × 100 two-stage link network*

Multistage switching

Multistage switching reduces the number of cross-points required to connect N input links to N output links. It also increases the number of paths between two endpoints, providing increased reliability. However, it introduces more control effort and internal blocking (inability to connect two stations because all possible paths are already in use). A 100×100 two-stage switch using 20 (10 each at the input and output) 10×10 cross-point switches is illustrated in Figure 2.17. The reader might have noticed that the two-stage network has only 2000 cross-points as against 10000 cross-points for a single 100×100 crossbar switch. For a two-stage network, let the occupancy of links be a, and the occupancy of the outgoing trunks be b. The, for an expander, the blocking $B_1 = a$. For a route switch, the blocking is:

$$B_2 = [1 - (1 - a)(1 - b)]^{g_2}, \tag{2.16.2}$$

where g_2 is the number of secondary switches.

Example 1 Find the grade of service when a total of 30 E is offered to the two-stage switching network of Figure 2.17 and the traffic is evenly distributed over the 10 outgoing routes.

Solution
The link and trunk occupancies are $a = b = 30/100 = 0.3$ E.

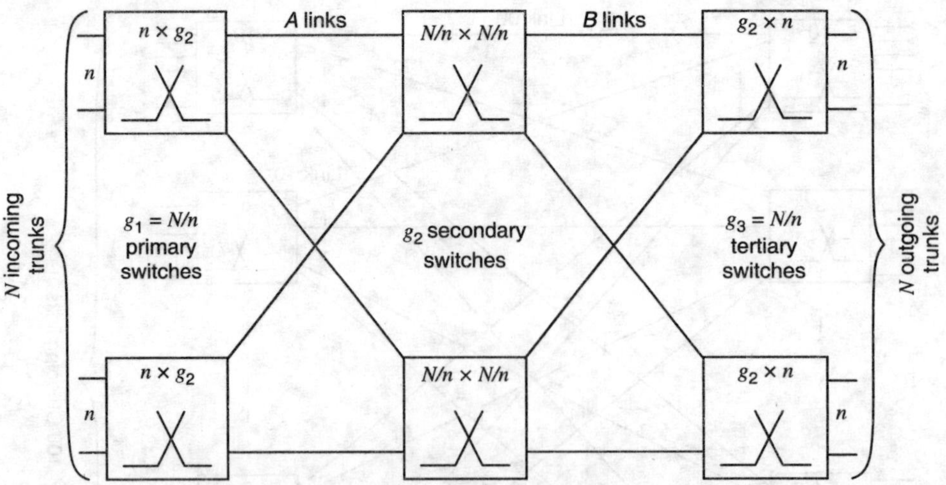

Figure 2.18 *An N × N three-stage link network*

$$B_2 = [1 - (1 - 0.3)(1 - 0.3)]^{10} = 0.51^{10} = 0.0012.$$

Example 2 Find the traffic capacity of this network if the grade of service is not to exceed 0.01.

Solution

$$B \leq 0.01 = [1 - (1 - a)^2]^{10}$$

$$1 - (1 - a)^2 \leq 0.01^{0.1} = 0.631$$

$$a \leq 0.39 \text{ and } A \leq 39 \text{ E}.$$

A fully interconnected three-stage network is shown in Figure 2.18. For this network, let the occupancy of A links be a, and occupancy of B links be b and the occupancy of outgoing trunks c. For an expander, probability that all g_2 independent paths are simultaneously blocked is $B_1 = [1 - (1 - a)(1 - b)]^{g_2}$, where g_2 is the number of secondary switches. For a route switch (i.e. a connection to any free trunk in a route having one trunk connected to each tertiary switch), the probability of blocking for a particular trunk is $1 - (1 - B_1)(1 - c)$. And, probability of simultaneous blocking for all g_3 independent paths is $B_2 = [B_1 + c(1 - B_1)]^{g_3}$, where g_3 is the number of tertiary switches.

Exercises

1. Calculate the time required to complete the dialling process when the number '290290' is dialed using a pulse dialler. (Answer: $\frac{1}{10} \times (2 + 9 + 10) \times 2 = 4.2$ seconds)
2. Repeat the above problem when DTMF dialler is used. (Answer: $100 \times 10^{-3} \times 6 = 0.6$ seconds, which will be same for all six-digit numbers)
3. Calculate the number of connections (links) required to connect N users in a *completely* connected network. (Answer: $\frac{N(N-1)}{2}$)

4. A DS1 digital signal encompasses 24 eight-bit PCM telephone channels time-division-multiplexed on a cable. One synchronising pulse per frame is provided, and the sampling frequency is 8 kHz. Determine the bit rate on the cable. (Answer: 1.544 Mbps)
5. A DS2 digital signal consists of four interleaved DS1 signals. One extra synchronisation byte is inserted every frame. Determine the bit rate of the DS2 system. (Answer: 6.24 Mbps)
6. Calculate the dc loop resistance of a telephone subscriber loop comprised of 5.6 km of AWG22 cable and a further 10 km of AWG19 cable. The round-trip loop resistances are $170\Omega/\text{mile}$ and $85\Omega/$ mile for AWG22 and AWG19 cables, respectively. (Answer: $\frac{5.6}{1.6} \times 170 + \frac{10}{1.6} \times 85 = 1126.25\Omega$.)
7. (a) Find the VNL of a trunk circuit involving 4000 km of microwave relay circuit with an average velocity of propagation of 299 mm/s. (Answer: 3.076 dB)
   (b) What would be the round-trip delay on the above circuit? (Answer: 26.76 ms).
   (c) How long would this circuit have to have echo suppressors? (Answer: every 3140 kms).
8. (a) A telephone circuit based on a microwave link traverses a 41500 km uplink path and a 42600 km downlink path. What is the round trip delay involved? (Answer: 560.67 ms)

   (b) Is an echo suppressor required? (Answer: VNL is 56.47 dB, yes).
9. Obtain the VNL needed for a 6000 km cable circuit on which the velocity of propagation is 95% of that of light. Does this circuit require an echo suppressor? (Answer: VNL=4.61 dB, round-trip delay=42.11 ms, yes)
10. Calculate the total number of cross-points in a three-stage network having ten 5×10 switches in the input and output stages and five 10×10 switches in the middle stage and is fully connected. If the link occupancies of the A and B links is 30 E and at the outgoing trunk is 45 E, obtain the grade of service. (Answer: $2 \times 5 \times 5 \times 10 + 5 \times 10 \times 10 = 1000$, $B_1 = (1 - (1 - .3)^2)^5 = 0.0345$, $B_2 = [0.0345 + 0.45(1 - 0.0345)]^{10} = 5 \times 10^{-4}$).
11. (a) Find the grade of service when a total of 30 E is offered to the two-stage switching network and the traffic is evenly distributed over the 10 outgoing routes.
   The link and trunk occupancies are $a = b = 30/100 = 0.3$ E.

   $$B_2 = [1 - (1 - 0.3)(1 - 0.3)]^{10} = 0.51^{10} = 0.0012.$$

   (b) Find the traffic capacity of this network if the grade of service is not to exceed 0.01.

   $$B \leq 0.01 = [1 - (1 - a)^2]^{10}$$

   $$1 - (1 - a)^2 \leq 0.01^{0.1} = 0.631$$

   $a \leq 0.39$ and $A \leq 39$ E

12. (a) Compare the grades of service provided by the two-stage networks when each operates as an expander and is offered 30 E of traffic. Assume the number of switches in the secondary stage is 5 and 10 respectively.

$a = b = 30/100 = 0.3$ E

For network (a): $B_1 = [1 - (1 - 0.3)(1 - 0.3)]^5 = 0.51^5 = 0.035$

For network (b): $B_1 = 0.51^{10} = 0.0012$

(b) What is the traffic capacity of each network if the required grade of service is 0.01?

For network (a):$[1 - (1 - a)^2]^5 = 0.01 \Rightarrow a = 0.224$

Total traffic capacity = 100 × 0.224 = 22.4 E

For network (b):$[1 - (1 - a)^2]^{10} = 0.01 \Rightarrow a = 0.393$

Total traffic capacity = 100 × 0.393 = 39.3 E.

13. Compute the time required to send a page as a facsimile if the data circuit has a speed of 9.6 kbps and the number of bytes in the scanned image is 160 kB. (Answer: $\frac{160\times8}{9.6} =$ 2.22 minutes).
14. An image file of size 1 MB is to be transmitted over a digital link with a speed of 19.2kbps. If a compression mechanism is employed before the image is transmitted, which would reduce the file size to 68%, calculate the time required to complete the transmission. (Answer: $\frac{1024\times8\times0.68}{19.2\times60} = 4.84$ minutes)
15. An electronic scanner produces 1.11 million pixels per document page and uses a run length code with Huffman coding to produce a 35 to 1 average data compression. If the maximum transmission bandwidth is 1 kHz, find the document transmission time. (Assuming 8 bits/pixel, $\frac{1.11\times10^6\times8}{35\times1000\times60} = 4.23$ minutes)
16. What frequencies will appear on a telephone loop when the * key is depressed on a touch-tone dialler? (Answer: 941 and 1209 Hz)

Chapter 3

Amplitude Modulation

3.1 Introduction

In the case of short-range systems, as in signal transmission in the telephone local loop or between the amplifier and the speakers of a sound system, the baseband signals are sent through wires. For longer distances, it is sometimes difficult to use wires as they require space and large quantities of material (usually copper) for doing so. Added to that, if the transmission is *point-to-point*, one must specify the location of every terminal. Mobile communication by wire is almost impossible, whereas if one resorts to radio communication, it becomes quite simple to move around. As mentioned in Chapter 1, the reasons for modulating (or varying in accordance with something else) the attributes (such as, amplitude, frequency, and phase) of a carrier wave are the following:

- Baseband has limited coverage.
- It is a less efficient means of transmission of information.
- Mobile communication is almost impossible with a wired baseband system.
- For efficiently radiating signals using an antenna, the spectrum has to be elevated to the higher side so as to reduce the antenna length.
- Multiplexing in the frequency domain is possible with modulation.

In amplitude modulation, we modulate (or vary in accordance with) the amplitude of a high-frequency carrier wave with a baseband modulating signal. If the carrier signal is represented by $A_c \sin \omega_c t$, and the modulating baseband is represented by $A_m \sin \omega_m t$, then the output of the amplitude modulator is given by

$$AM_{out} = A_c\{1 + A_m \sin \omega_m t\} \sin \omega_c t. \tag{3.1.1}$$

A simple set of MATLAB code to generate and plot an amplitide modulated (AM) signal is give below along with the results in Figure 3.1.

```
%% An m-file to generate and plot AM (A3E) signal.%%%
clf;
t=[0:.0001:.1];
fc=1000;  % carrier frequency.
fm=50;    % modulating frequency.
wc=2*pi*fc*t;
```

```
wm=2*pi*fm*t;
subplot(311),plot(t,sin(wm)),...
xlabel('Time'),ylabel('sin\omega_m t'),title('Modulating
Signal'); subplot(312),plot(t,sin(wc)),...
xlabel('Time'),ylabel('sin\omega_c t'),title('Carrier
Signal'); subplot(313),plot(t,(1+sin(wm)).*sin(wc)),...
xlabel('Time'), ylabel('AM_{out}'),...
title('Amplitude Modulated Output Signal (A3E)');
%% End of Program am.m %%%%
```

More or less identical results can be obtained using the MATLAB built-in function 'modulate(). The commands to generate the plot given in Figure 3.1 is as follows:

```
t=[0:.0001:.1];
fc=1000; % Carrier frequency
fs=2500;% Sampling frequency, >= 2*fc+fm
fm=50; % modulating signal frequency.
wm=2*pi*fm*t;
[y T]=modulate(sin(wm),fc,fs,'amdsb-tc');
plot(T,y);
xlabel('Time'), ylabel('AM_{out}'),...
title('Amplitude Modulated Output Signal (A3E)');
```

3.1.1 Commercial AM broadcast band

The commercial AM band of frequencies was opened for the intended use of voice communications only. Therefore, 10 kHz was allocated to each of the channels in the AM band. The lowest frequency in the band is identified as channel 1, having a carrier frequency of 540 kHz, and it extends from 535 to 545 kHz. The AM band can incorporate 107 channels altogether. However, no metropolitan area can support 107 AM broadcast stations because of the financial burden. Hence, it is common to find a gap of 30 to 70 kHz between assigned AM stations in any locality. Because of this wide channel separation, many AM broadcast stations extend their audio frequencies to provide better fidelity.

3.2 Classification of Various AM Systems

If we look at the spectrum of the signal, we can find that it contains the carrier frequency (f_c) and two sidebands ($f_c + f_m$ and $f_c - f_m$). As far as the information transfer is concerned, the carrier frequency is redundant. *This is do to with the fact that the carrier component remains constant in amplitude and frequency no matter what the modulating signal is.* So also it is with one of the sidebands. This means that we need to transmit only one of the sideband (that contains the baseband frequencies), and we will still be able to retrieve the information at the receiver. Although this results in a saving on the bandwidth, it complicates the transmitter and/or receiver design. Therefore, this single sideband (a.k.a. SSB transmission) is rarely used in practice. However, SSB has an advantage in systems where transmitted power needs to be

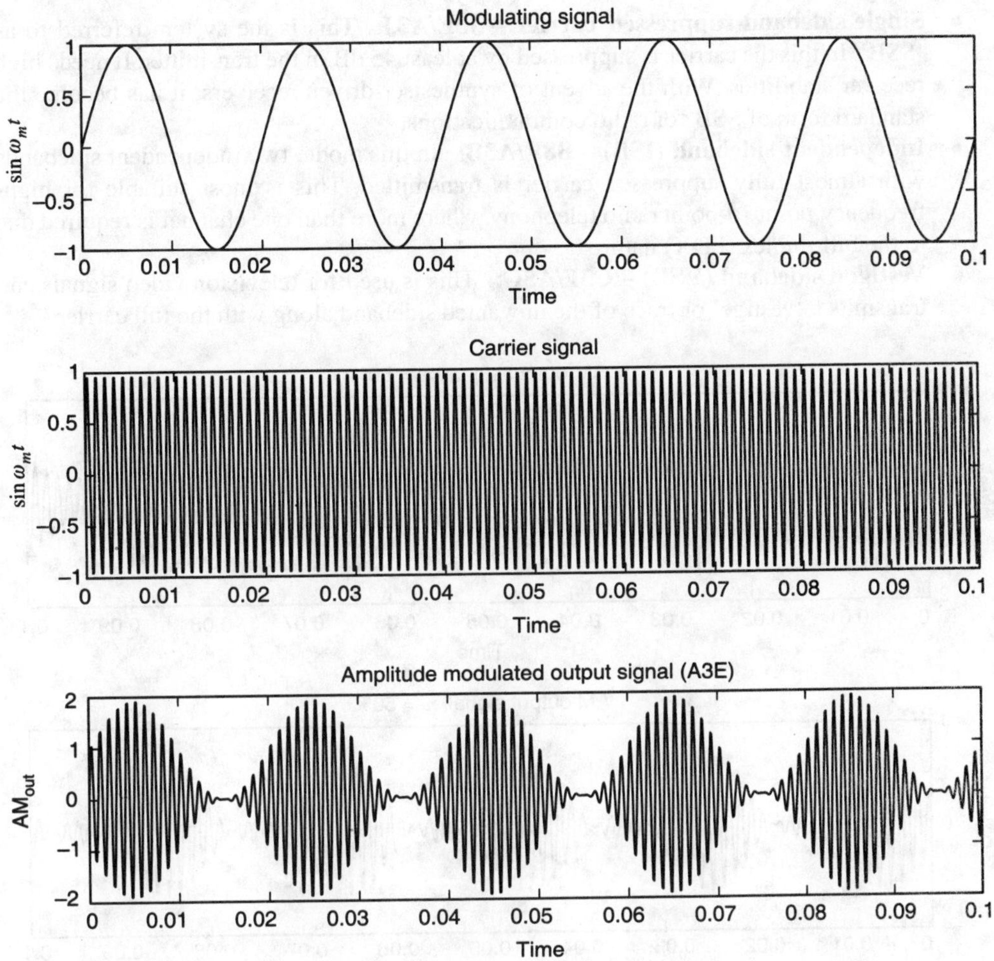

Figure 3.1 *Amplitude modulated output (A3E) signal*

kept low (for example, mobile systems), as we save on the power of the carrier and that of the suppressed sideband. Practically, several variants of AM are used, which are described below.

- **Standard AM – A3E/A3**: It is named as A3E by the International Telecommunications Union (ITU) Radio regulations. It is used mainly for broadcasting and not for anything else now.
- **Single sideband reduced carrier – R3E/A3A**: This is the R3E system as per ITU specifications and is basically a *pilot-carrier system.* An attenuated carrier signal is reinserted to the SSB signal to facilitate receiver tuning and demodulation. It is mainly used for maritime mobile distress communication (a.k.a. SOLAS – Safety of Life At Sea) at 2182 kHz.
- **Single sideband full carrier – H3E/A3H**: This is compatible with the AM broadcasting system, with A3E receivers. A distortion not exceeding 5% is claimed for H3E transmission received by the A3E receiver.

- **Single sideband suppressed carrier – J3E/A3J**: This is the system referred to as 'SSB'. In this the carrier is suppressed by at least 45 dB in the transmitter. It needs high receiver stabilities. With the advent of synthesiser-driven receivers, it has become the standard form of SSB for radio communications.
- **Independent sideband (ISB) – B8E/A3B**: In this mode, two independent sidebands with almost fully suppressed carrier is transmitted. This is most suitable for high-frequency point-to-point radio telephony, where more than one channel is required due to the full-duplex (FDX) nature.
- **Vestigial sideband (VSB) – C3F/A5C**: This is used for television video signals and transmits a 'vestige' or trace of the unwanted sideband along with the full carrier.

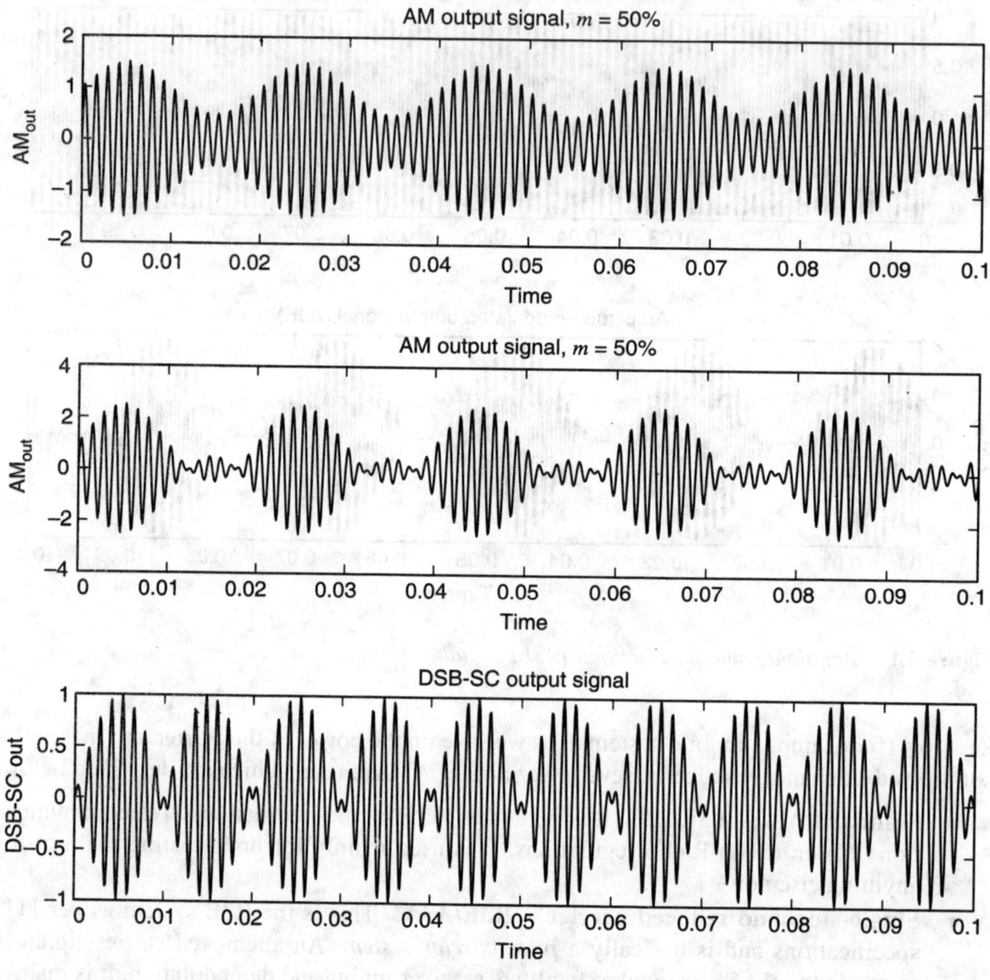

Figure 3.2 *AM output for various values of modulation index*

3.3 Frequency Spectrum

When a carrier is amplitude modulated, the instantaneous modulating voltage variations are superimposed onto the carrier amplitude. This is illustrated in Figure 3.2, from which it is also evident that distortion will occur if V_m is greater than V_c. This is due to overdriving the amplifier stage. Now we define the ratio $\frac{V_m}{V_c}$ as the *modulation index, m.*

$$m = \frac{V_m}{V_c} \tag{3.3.2}$$

The MATLAB .m file for the above plots given in Figure 3.2 is as follows:

```
%% An m-file to generate and plot AM signal for Various
Values of m.clf;
t=[0:.00001:.01];
fc=8000;   % carrier frequency.
fm=500;    % modulating frequency.
wc=2*pi*fc*t;
wm=2*pi*fm*t;
subplot(311),plot(t, (1+.5*sin(wm)).*sin(wc)),...
xlabel('Time'),ylabel('AM_{out}'),title('AM Output Signal,
m=50%'); subplot(312),plot(t,(1+1.5*sin(wm)).*sin(wc)),...
xlabel('Time'),ylabel('AM_{out}'),title('AM Output Signal,
m=150%'); subplot(313),plot(t,sin(wm).*sin(wc)),...
xlabel('Time'),ylabel('DSBSC out'),title('DSBSC
Output Signal');
%% end of am2.m
```

As the modulation index has a numerical value between 0 and 1, it is expressed as a percentage and is called *percentage modulation.* Figure 3.3 gives another insight into the relationship between V_{max} and V_{min} of the AM wave and the modulation index, m.

It can be shown that

$$V_m = \frac{V_{max} - V_{min}}{2} \quad \text{and}$$

$$\begin{aligned} V_c &= V_{max} - V_m = V_{max} - \frac{V_{max} - V_{min}}{2} \\ &= \frac{V_{max} + V_{min}}{2}. \end{aligned} \tag{3.3.3}$$

$$\begin{aligned} \text{Therefore, } m &= \frac{V_m}{V_c} = \frac{(V_{max} - V_{min})/2}{(V_{max} + V_{min})/2} \\ &= \frac{V_{max} - V_{min}}{V_{max} + V_{min}}. \end{aligned} \tag{3.3.4}$$

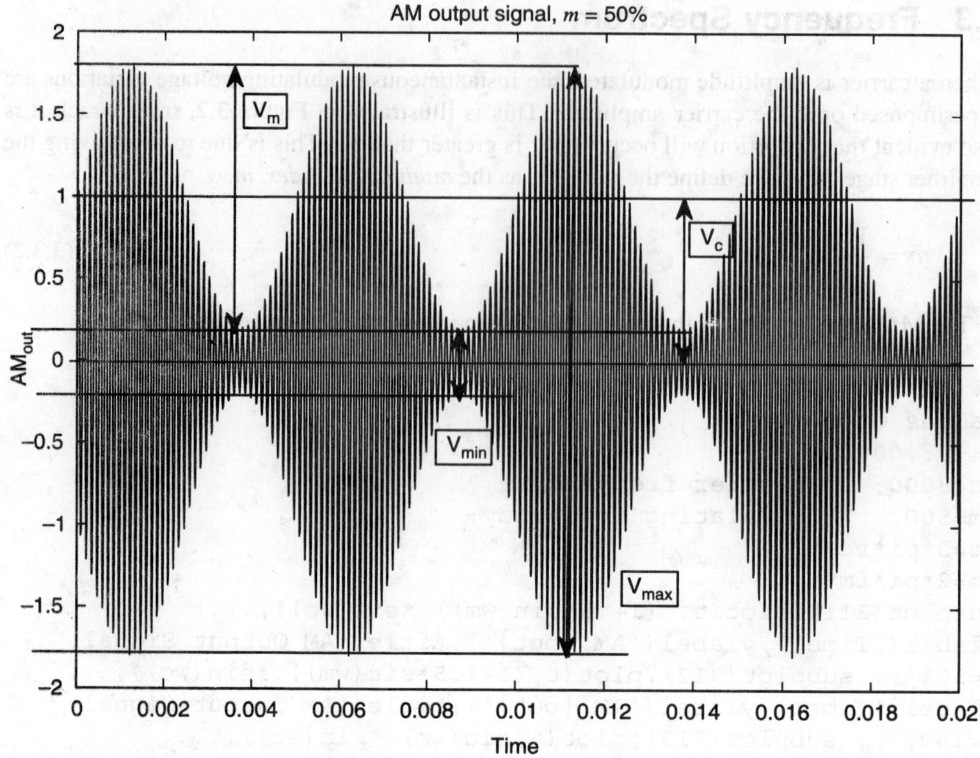

Figure 3.3 *Amplitude modulated wave*

The above equation is very useful for computing modulation index from an oscilloscope display. Now, the amplitude of the modulated signal is given by

$$A = V_c + V_m \sin \omega_m t \tag{3.3.5}$$

$$= V_c + m V_c \sin \omega_m t \tag{3.3.6}$$

$$= V_c(1 + m \sin \omega_m t). \tag{3.3.7}$$

The instantaneous voltage of the resulting AM wave is:

$$\text{AM}_{\text{out}} = \text{A} \sin \theta = A \sin \omega_c t \tag{3.3.8}$$

$$= V_c(1 + m \sin \omega_m t) \sin \omega_c t. \tag{3.3.9}$$

Expanding (3.3.9), we have:

$$\text{AM}_{\text{out}} = V_c \sin \omega_c t + m V_c \sin \omega_c t \sin \omega_m t \tag{3.3.10}$$

$$= V_c \sin \omega_c t + \frac{m V_c}{2} \cos (\omega_c - \omega_m)t - \frac{m V_c}{2} \cos (\omega_c + \omega_m)t. \tag{3.3.11}$$

It is evident from the above equation that the AM output signal contains the carrier signal and two sidebands. The frequency of the lower sideband (LSB) is $f_c - f_m$ and that of the upper sideband (USB) is $f_c + f_m$. The bandwidth required for the amplitude modulation is thus *twice* that of the modulating signal. The frequency spectrum of the AM wave is illustrated in Figure 3.4.

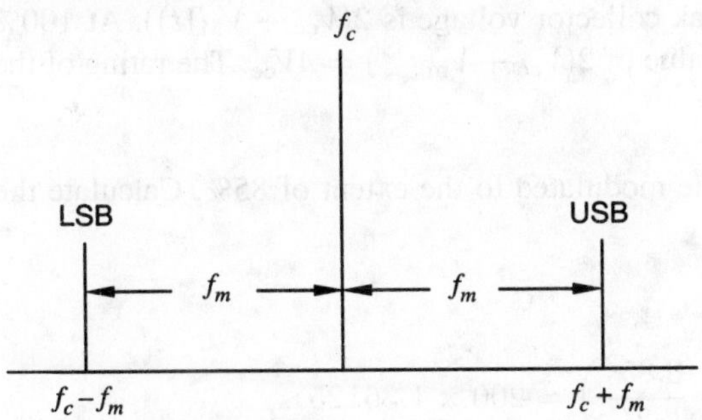

Figure 3.4 *The frequency spectrum of the AM wave*

3.4 Power Relations of AM Components

It is clear from earlier equations for the AM output signal that the carrier component of the modulated signal is same as that of the unmodulated carrier. The modulated wave contains extra energy because of the sidebands. It is also clear that total power in the modulated wave will depend on the modulation index. The total power in the modulated wave will be

$$P_t = \frac{{V_{Carr}}^2}{R} + \frac{{V_{LSB}}^2}{R} + \frac{{V_{USB}}^2}{R} \text{ (rms)}, \tag{3.4.12}$$

where all the voltages are RMS values ($= \frac{1}{\sqrt{2}} \times V_{peak}$), and R is the antenna resistance in which the power is dissipated. The unmodulated carrier power can be given as

$$P_c = \frac{{V_{Carr}}^2}{R} = \frac{(V_c/\sqrt{2})^2}{R} = \frac{{V_c}^2}{2R}. \tag{3.4.13}$$

Likewise,

$$P_{LSB} = P_{USB} = \frac{V_{SB}^2}{R} = \left(\frac{mV_c}{2\sqrt{2}}\right)^2 \frac{1}{R} = \frac{m^2 {V_c}^2}{8R} = \frac{m^2}{4}\frac{{V_c}^2}{2R}. \tag{3.4.14}$$

Substituting in the previous equations, we have:

$$P_t = \frac{V_c^2}{2R} + \frac{m^2}{4}\frac{{V_c}^2}{2R} + \frac{m^2}{4}\frac{{V_c}^2}{2R} = P_c + \frac{m^2}{4}P_c + \frac{m^2}{4}P_c.$$

$$\frac{P_t}{P_c} = 1 + \frac{m^2}{2}. \tag{3.4.15}$$

If P_{SB} is the total power in sidebands and P_{mod} is the average supplied by the modulator, then the efficiency of conversion, η, is given by

$$\eta = \frac{P_{SB}}{P_{mod}}. \tag{3.4.16}$$

It should also be noted that the peak-to-peak collector voltage is $2(V_{cc} + V_M(t))$. At 100% modulation, this voltage can reach a peak value of $2(V_{cc} + V_{m_{max}}) = 4V_{cc}$. The rating of the transistor must take this into account.

Example 3.1 A 200 W carrier is amplitude modulated to the extent of 85%. Calculate the total power in the modulated wave.

Solution

$$\begin{aligned} P_t &= P_c\left(1 + \frac{m^2}{2}\right) = 200\left(1 + \frac{0.85^2}{2}\right) = 200 \times 1.36125 \\ &= 272.25 \text{ W}. \end{aligned}$$

Example 3.2 The current of the transmitting antenna of an AM transmitter is 10 A when only the carrier is sent, but it rises to 12 A when it is modulated by a single sine wave.

1. Find the modulation index.
2. Determine the antenna current when the modulation index is changed to 90%.

Solution

$$\begin{aligned} \frac{P_t}{P_c} &= 1 + \frac{m^2}{2} = \frac{I_t{}^2 R}{I_c{}^2 R} = \left(\frac{I_t}{I_c}\right)^2 \\ m &= \sqrt{2\left[\left(\frac{I_t}{I_c}\right)^2 - 1\right]} \\ &= \sqrt{2\left[\left(\frac{12}{10}\right)^2 - 1\right]} = \sqrt{2\,[1.44 - 1]} \\ &= 93.81\%. \end{aligned}$$

For the second part we have

$$I_t = I_c\sqrt{1 + \frac{m^2}{2}} = 10\sqrt{1 + \frac{0.9^2}{2}} = 11.85 \text{ A}.$$

3.4.1 Modulation by several signals

In practice, the carrier is modulated by a number of sine wave signals at a time. And, most often, a *real-world signal* such as the audio will consist of a number of sine waves, which may be harmonically related.[1] In any case, we are interested in obtaining the overall modulation index when the carrier is amplitude modulated by several signals of RMS voltage values V_1, V_2, V_3, etc. Now the resultant RMS voltage is given by

$$V_t = \sqrt{V_1^2 + V_2^2 + V_3^2 + \dots} .$$

Dividing both sides by V_c, the RMS voltage of the carrier, we get

$$\begin{aligned} \frac{V_t}{V_c} &= \frac{\sqrt{V_1^2 + V_2^2 + V_3^2 + \dots}}{V_c} \\ &= \sqrt{\frac{V_1^2}{V_c^2} + \frac{V_2^2}{V_c^2} + \frac{V_3^2}{V_c^2} + \dots} \\ &= \sqrt{m_1^2 + m_2^2 + m_3^2 + \dots} . \end{aligned} \quad (3.4.17)$$

Example 3.3 An AM broadcast transmitter radiates 10 kW of power when unmodulated and 12.5 kW when modulated by a sine wave.

1. Calculate the modulation index/percentage of modulation.
2. If another sine wave simultaneously modulates the carrier to a depth of 50%, determine the total power transmitted.

Solution

$$\begin{aligned} \frac{m^2}{2} = \frac{P_t}{P_c} - 1 = \frac{12.5}{10} - 1 &= 1.25 - 1 = 0.25 \\ m^2 &= 0.25 \times 2 = 0.5 \\ m &= \sqrt{0.5} = 0.7071. \end{aligned}$$

For the second part, the total modulation index is

$$\begin{aligned} m_t &= \sqrt{m_1^2 + m_2^2} = \sqrt{0.7071^2 + 0.5^2} = \sqrt{0.5 + 0.25} = \sqrt{0.75} = 86.60\%. \\ P_t &= P_c\left(1 + \frac{m_t^2}{2}\right) = 10\left(1 + \frac{0.806^2}{2}\right) = 10\left(1 + \frac{0.75}{2}\right) = 10(1 + 0.375) \\ &= 13.75 \text{ kW}. \end{aligned}$$

3.5 DSB-SC Modulation

A double sideband suppressed carrier (DSB-SC) signal has no dc component added to the signal. The modulated signal is the product of the modulating signal $V_m \sin \omega_m t$ and the carrier, $V_c \sin \omega_c t$ with the result that

[1] A square wave, for example, can be synthesised by summing several harmonically related sine waves.

$$\text{DSB-SC}_{\text{out}} = V_c \sin \omega_c t \, V_m \sin \omega_m t = m V_c^2 \sin \omega_c t \sin \omega_m t \quad (3.5.18)$$

$$= m V_c^2 \{\cos (\omega_c - \omega_m)t - \cos (\omega_c + \omega_m)t\}. \quad (3.5.19)$$

Comparing this with equation (3.3.11) for the case of AM with carrier, we see that equation (3.5.19) has no carrier frequency term. Also, refer to the last segment of Figure 3.2 as well. Notice that there is no envelope of the original modulating signal in the modulated signal as there is in double sideband AM with carrier. With no component at the carrier frequency, the transmission power requirements are lower than that for AM with carrier.

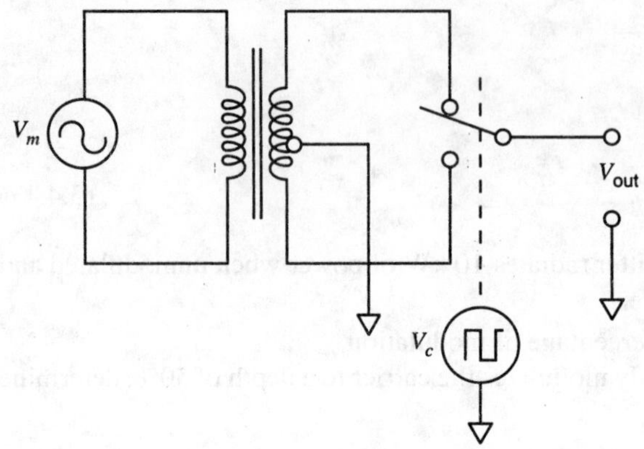

Figure 3.5 *Switching DSB-SC modulator*

The circuit shown in Figure 3.5 generates DSB-SC output. The function of the voltage v_c is to drive the switch to its two alternate positions at the carrier frequency rate. In the output there would be a square wave carrier instead of the usual sinusoid; however, the higher harmonics of the square wave carrier could readily be eliminated by filtering.

3.5.1 Balanced modulator

A relatively straightforward method to generate DSB-SC modulation is to use two conventional AM modulators arranged in the configuration illustrated in Figure 3.6. Any *square law* AM modulators can be used in this system. But care must be given to choose modulators with approximately identical characteristics so that the carrier component cancels out at the summing amplifier.

3.5.2 Ring modulator

Another type of modulator for generating a DSB-SC AM signal is the *ring modulator*, illustrated in Figure 3.7. The switching of the diodes is controlled by a square wave of frequency f_c, denoted as $c(t)$, which is applied to the centre tap of the two transformers. When $c(t) > 0$, the top and bottom diodes conduct (i.e., D_1 and D_3), while the other diodes, viz., D_2 and D_4, are off. In this case, the message signal $m(t)$ is multiplied by +1. When $c(t) < 0$, the diodes D_2 and D_4 conduct, while the other two diodes are off, and the signal is multiplied by

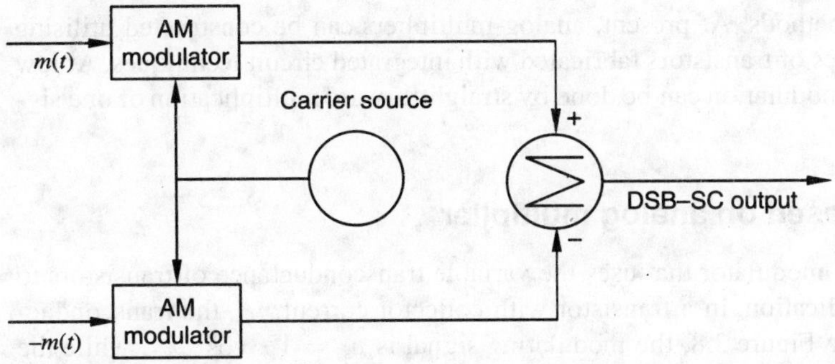

Figure 3.6 *Block diagram of a balanced modulator*

−1. Consequently, the operation of the ring modulator can be described mathematically as a multiplier of $m(t)$ by the square wave carrier $c(t)$, i.e.,

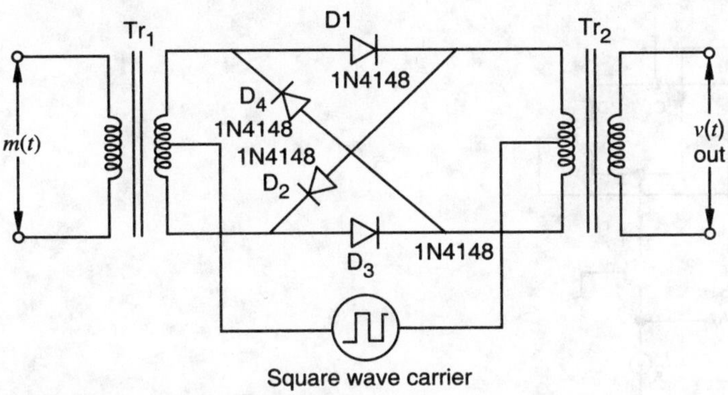

Figure 3.7 *Ring modulator*

$$v_{\text{out}}(t) = m(t)c(t). \tag{3.5.20}$$

Since $c(t)$ is a periodic function, it is represented by the Fourier series:

$$c(t) = \frac{4}{\pi}\sum_{n=1}^{\infty}\frac{(-1)^{n-1}}{2n-1}\cos\left[2\pi f_{\text{c}}(2n-1)t\right]. \tag{3.5.21}$$

Thus the DSB-SC AM signal $u(t)$ is obtained by passing $v_{\text{out}}(t)$ through a bandpass filter with centre frequency f_{c} and bandwidth $2W$.

3.6 Amplitude Modulator Circuits

As was emphasised in the previous sections, the modulation process is essentially multiplication of one signal with another. In the earlier days of the electronic practice, analog multipliers were relatively inaccurate and had limited capability, and so methods were employed using nonlinear circuit methods. At present, analog multipliers can be constructed utilising the matched characteristics of transistors fabricated with integrated circuit techniques. At low power levels, amplitude modulation can be done by straightforward multiplication of one signal with another.

3.6.1 Modulator based on analog multiplier

Figure 3.8 shows an AM modulator that uses the variable transconductance of transistors to effectively obtain multiplication. In a transistor with collector current, i_C, the transconductance is $\frac{qi_C}{kT}$. Referring to Figure 3.8, the modulating signal is $v_s = V_S \cos \omega_{st}$, while the carrier signal is $v_c = V_C \cos \omega_c t$. The modulating signal causes the currents i_1 and i_2 to have the values and defining transconductances g_{m1} and g_{m2}.

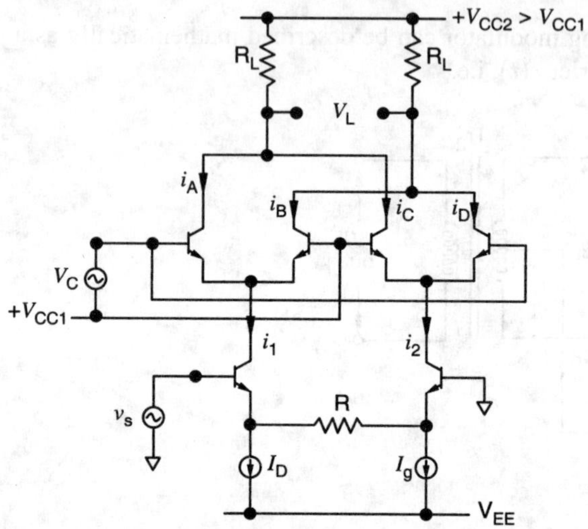

Figure 3.8 *A DSB-SC modulator using differential amplifier transistor configuration. The four transistors on the upper part of the configuration form the two differential amplifier pairs which act as the transconductance amplifiers. The modulating signal decides the transconductance values.*

$$i_1 = I_0 + \frac{v_s}{R}, \quad i_2 = I_0 - \frac{v_s}{R} \tag{3.6.22}$$

$$g_{m1} = \frac{qi_1}{kT}, \quad g_{m2} = \frac{qi_2}{kT} \tag{3.6.23}$$

$$g_{m1} - g_{m2} = \frac{q}{kT}(i_1 - i_2) = \frac{2q}{kT}\frac{v_s}{R}. \tag{3.6.24}$$

We have

$$i_A = \frac{I_0}{2} + \frac{g_{m1}}{2}v_c, \quad i_B = \frac{I_0}{2} - \frac{g_{m1}}{2}v_c, \quad i_C = \frac{I_0}{2} - \frac{g_{m2}}{2}v_c, \tag{3.6.25}$$

$$i_D = \frac{I_0}{2} + \frac{g_{m2}}{2}v_c.$$

Since

$$v_L = (i_A - i_B + i_C - i_D)R_L = (g_{m1} - g_{m2})v_C R_L,$$

$$v_L = \frac{2qV_S}{kT}\frac{V_C}{R} R_L \cos\omega_s t \cos\omega_c t \qquad (3.6.26)$$

which is a DSB-SC AM signal. To generate a double sideband AM signal with carrier (A3E), two dc current sources can be unbalanced (which is not shown). If the two currents are I_{01} and I_{02}, the resulting modulated signal is

$$v_L = \left[\frac{q(I_{01} - I_{02})}{kT} + \frac{2qV_S}{kTR}\cos\omega_s t\right] R_L V_C \cos\omega_c t \qquad (3.6.27)$$

which is an AM double sideband signal with carrier (A3E), where

$$m = \frac{2V_S}{(I_{01} - I_{02})R}, \quad \text{and carrier amplitude} = \frac{q(I_{01} - I_{02})}{kT} R_L V_C. \qquad (3.6.28)$$

3.6.2 Transistorised modulators

There can be two types of transistorised modulators, high level and low level, depending on where the modulating signal is applied. If we apply the modulating signal to the collector, high-level modulation results, otherwise low level. Figure 3.9 shows a typical emitter modulation circuit (low-level modulation). Here we will be modulating a low-power LC (*Colpitts oscillator*), where the carrier frequency is decided by L_1, C_1, and C_2. A high-level modulator

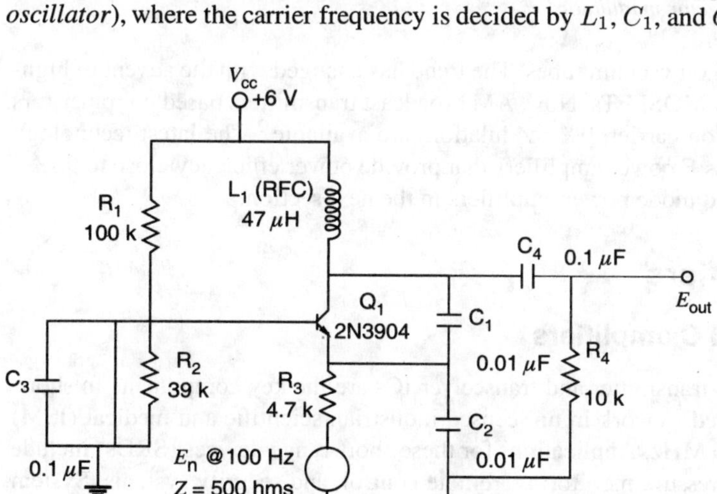

Figure 3.9 *Low-level emitter modulator*

is illustrated in Figure 3.10. The circuit is known as collector modulator or *Heising modulator*, named after its inventor. As illustrated in Figure 3.10, the amplifier is a biased class C and the output RF signal amplitude is modulated by causing the collector supply voltage to vary with the modulation voltage. This method is capable of producing nearly 100% modulation with reasonable linearity. It should be noted that, in the past, almost *all* high-power broadcast

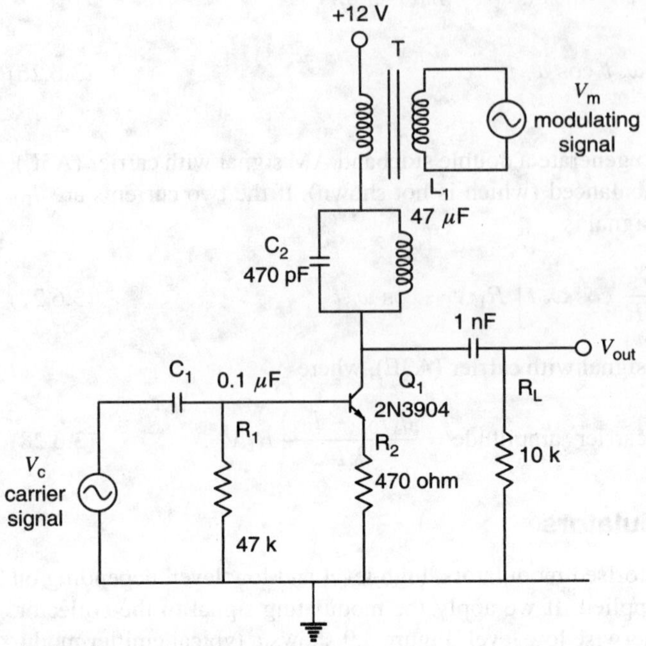

Figure 3.10 *High-level collector modulator*

AM transmitters were based on vacuum tubes. The trend has changed with the advent of high-power transistors and power MOSFETs. Now, AM broadcast transmitters based on transistors up to about 10 kW (power on carrier, 0% modulation) are available.[2] The latest technology also includes class E or class F power amplifiers that provide power efficiency close to 90%.[3] We will discuss the switched mode power amplifiers in the next section.

3.7 Power Amplifiers

3.7.1 Class A, B and C amplifiers

Low-cost ASK and/or FSK transmitter and transceiver ICs are the key components in short-range radio systems designed to work in unlicensed industrial, scientific and medical (ISM) bands between 300 and 450 MHz. Applications for these short-range devices (SRDs) include remote keyless entry, tire pressure monitoring, remote control, and security systems. System design, which addresses RF link budgeting, antenna design, battery life, regulatory concerns and other issues, often involves a trade-off between the output power and the current drain of the transmitter.

[2]In India, Bharat Electronics Limited, Bangalore, designs and manufactures AM and FM broadcast transmitters for All India Radio. Refer their product catalogue for details.
[3]*www.bel-india.com*

A class A amplifier is characterised by a combination of bias point and signal level where the device's average current drain does not change with the magnitude of the input signal. In Figure 3.11, M_1 can be assumed to be a current source of magnitude I_{DC}. It is well known that the impedance for maximum output power is

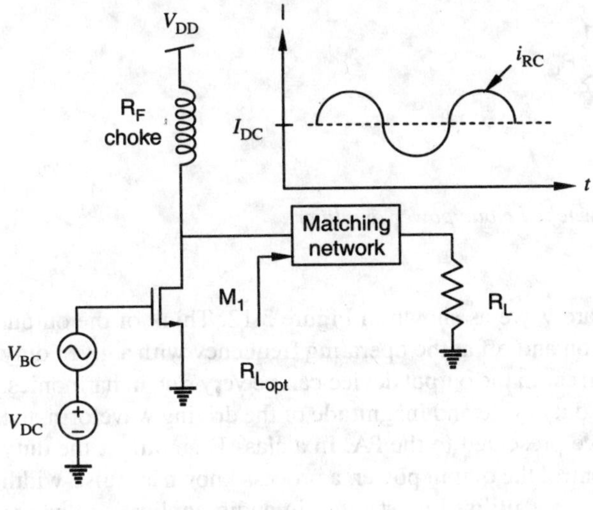

Figure 3.11 *Schematic diagram of a class A power amplifier*

$$RL_{opt} = V_{DD}/I_{DC}. \tag{3.7.29}$$

The maximum output power is defined as

$$P_{out}(\max) = V_{DD} \times I_{DC}. \tag{3.7.30}$$

Therefore, the peak efficiency is 50%. This analysis assumes that the drain voltage of M1 can swing to ground while still maintaining a bias current of I_{DC}. Operation in the triode region limits the practical efficiency of class A CMOS power amplifiers to less than 40%. As this analysis implies, the bias current of a class A amplifier must be changed to maintain a reasonably high efficiency for different output-power levels for a given supply voltage. Class A amplifiers are most amenable to modulation schemes where linear amplification of the input signal is important, because the bias point does not change with the magnitude of the input signal.

Class B and class C amplifiers, however, offer higher efficiencies than the class A amplifier, but usually at reduced output-power levels and with more distortion.

3.7.2 Class D, E, and F amplifiers

In contrast to class A, B, and C amplifiers, where operation in the triode region should be avoided, the class D, E, and F CMOS amplifiers rely on operation in the triode region for optimum efficiency and output power. These amplifiers are often called 'switched-mode' amplifiers and are typically used in ISM-band transmitters and transceivers because of their inherent high-efficiency operation at low voltages. In a switching-mode amplifier, the output

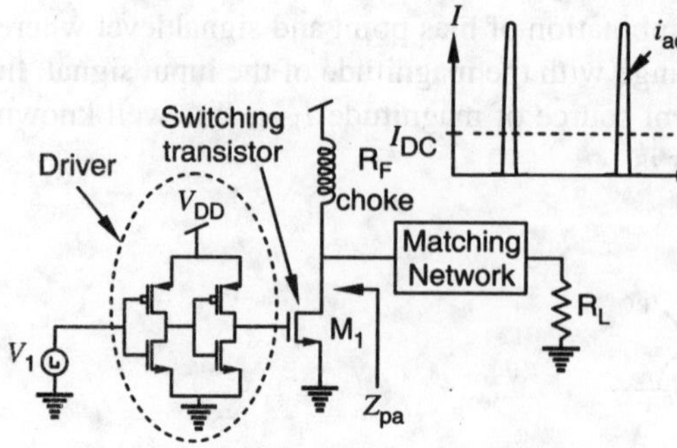

Figure 3.12 *Schematic diagram of a switched mode power amplifier*

device is driven by a large-signal square wave as shown in Figure 3.12. Think of the output transistor as a resistor that is switched on and off at the operating frequency with a given duty cycle. As Figure 3.12 indicates, the current in the output device can be very rich in harmonics. This harmonic content depends on the duty cycle and magnitude of the driving waveform, the FET 'on' resistance, and the impedance presented to the PA. In a class D amplifier, the duty cycle of the input signal is varied to control the output power, a process known as pulse-width modulation (PWM). Class D amplifiers are utilised most often in audio applications where the power delivered by the amplifier changes constantly.

In a class E amplifier, the duty cycle of the input signal is fixed. The matching network is designed to minimise the voltage at the drain terminal of the switching while the switch is on. By minimising the voltage across the output device while the output device draws current, one can minimise the power dissipated by the switching device and, therefore, maximise PA efficiency.

Similar to a class E amplifier, a class F amplifier requires special attention for the harmonic impedances in the design of the matching network in order to enhance efficiency. In general, the matching circuits for class F amplifiers are more complex because of the design constraints placed on harmonic impedances.

A traditional power amplifier classification/comparison is given in Table 3.1.

Table 3.1 *Traditional PA classification*

Class	Modes	Conduction angle (%)	Output power	Maximum efficiency(%)	Gain	Linearity
A		100	Moderate	50	Large	Good
B	Current source	50	Moderate	78.5%	Moderate	Moderate
C		<50	Small	100	Small	Poor
D		50	Large	100	Small	Poor
E	Switch	50	Large	100	Small	Poor
F		50	Large	100	Small	Poor

3.7.3 Summary

In summary, some of the important highlights and characteristics of ISM-band switched-mode amplifiers are:

- Switched-mode amplifiers rely on operation in the triode region for optimum efficiency and output power at low supply voltages. This contrasts with class A, B, and C amplifiers, where operation in the triode region is to be avoided.
- All CMOS ISM switching-mode amplifiers provide an open-drain PA output. The user designs the matching network that provides the desired output power level, current drain, and harmonic performance. This flexibility allows the user to tailor the RF power/current drain trade-off while maintaining high efficiency, which is the key to maximising battery life for a particular application.
- To maximise the efficiency of a switching-mode amplifier, the switch must be turned on only near a minimum in the voltage waveform. Minimising the imaginary component of the impedance presented to the PA at the operating frequency (including the parasitic capacitances of the device, package, and board) satisfies this requirement.
- Depending on the particular PA load impedance, a current drain minimum can occur at resonance. This knowledge is useful for verifying that a given network has been optimised for a particular operating frequency and load.

3.8 Dynamic Carrier Control

It is easily shown that we can save transmitted power in AM transmitters if the carrier power is reduced. This will also result in longer life of the vacuum tubes or transistors used in the power amplifier stages. The technique is to keep the depth of modulation close to 100% dynamically by reducing the carrier level whenever the modulating signal level drops. Although this would reduce the range of reception, for all practical purposes the change is imperceptible. Therefore, dynamic carrier control is employed in all modern AM transmitters.

3.9 Demodulator Circuits

There are basically two techniques used to demodulate AM signals: coherent detectors and non-coherent detectors. The simplest among detectors is the non-coherent detector.

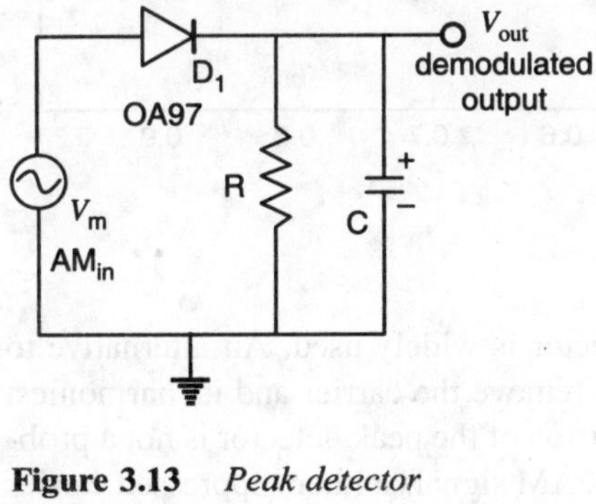

Figure 3.13 *Peak detector*

3.9.1 The non-coherent detector

Non-coherent (asynchronous) AM detectors employ either

- rectifier demodulation,
- envelope detection,
- square-law detection.

A simple form of non-coherent detector is the peak detector or the envelope detector. Figure 3.13 shows the circuit. Output of the peak detector before filtering (simulated) is illustrated in Figure 3.14. As it is evident, the capacitor charges on the rising part of the waveform and discharges on the falling part of the waveform. Normally, as the level of the signal is low, a Germanium diode like the OA79 is used in place of D_1 in Figure 3.13. The two types of distortion that appear in the demodulated output are carrier ripple on the rising edge and, possibly, an exponential decay instead of a sinusoidal variation on the falling edge if the RC

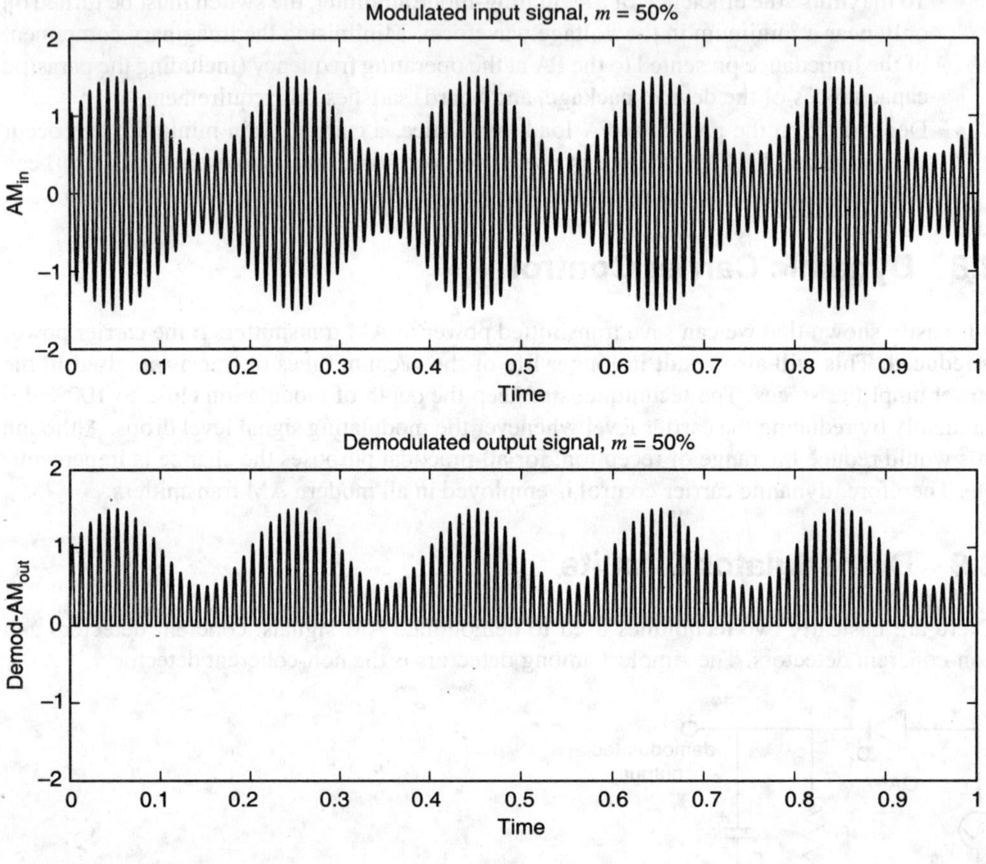

Figure 3.14 *Output of the peak detector*

time constant is too large. Nevertheless, the peak detector is widely used. An alternative to peak detection is rectification with lowpass filtering to remove the carrier and its harmonics. In this demodulation method the failure to follow distortion of the peak detector is not a problem. If the carrier signal is available at the demodulator, AM signals, either suppressed carrier or with carrier, can be demodulated by multiplying with carrier signal and lowpass filtering.

3.9.2 The coherent detector

The coherent detector is a synchronous detector where the frequencies generated in the receiver and used for demodulation are synchronised to oscillator frequencies generated in the transmitter (the receiver must have some means of recovering the received carrier and synchronising to it). In coherent demodulation, a locally generated carrier is synchronised to the transmitter's carrier frequency. *This technique is rarely used in practice.* The block diagram of a coherent AM detector is shown in Figure 3.15.

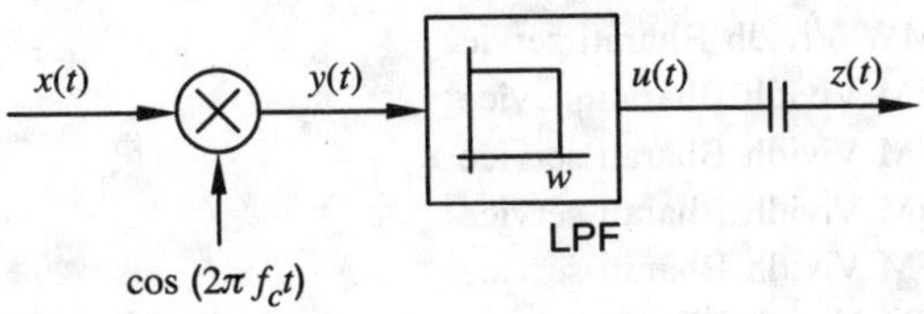

Figure 3.15 *Block diagram of coherent AM detector*

3.10 Radio Broadcasting in India

Sound broadcasting started in India in 1927 with the proliferation of private radio clubs. The operations of All India Radio (AIR) began formally in 1936 as a government organisation, with clear objectives to inform, educate and entertain the masses.

When India attained Independence in 1947, AIR had a network of just six stations and a complement of 18 transmitters. The coverage was 2.5% of the area and just 11% of the population. Rapid expansion of the network took place in the post-Independence era.

AIR today has a network of 214 broadcasting centres with 143 medium frequency (medium wave, MW), 54 high frequency (short wave, SW), and 139 FM transmitters. The coverage is 91.37% of the area, serving 99.13% of the people in the largest democracy of the world. AIR covers 24 languages and 146 dialects in home services. In external services it covers 27 languages, 17 national and 10 foreign languages. The frequency allocations of major AIR stations are as shown in Table 3.2.

Table 3.2 *Frequency allocations of AIR stations in India*

Sl.No.	Name of Station	Frequency	Type of service
1	Chennai C	783 kHz	MW Vividh Bharati service
2	Cuttack B	1314 kHz	MW Vividh Bharati service
3	Delhi C	1368 kHz	MW Vividh Bharati service
4	Dharwad B	1350 kHz	MW Vividh Bharati service
5	Jaipur B	1269 kHz	MW Vividh Bharati service
6	Jalandhar C	1350 kHz	MW Vividh Bharati service
7	Kanpur	1449 kHz	MW Vividh Bharati service
8	Kolkata C	1323 kHz	MW Vividh Bharati service

(Continuous)

Tale 3.2 *Frequency allocations . . .* *(Continued)*

Sl.No.	Name of Station	Frequency	Type of service
9	Kozhikode B	1431 kHz	MW Vividh Bharati service
10	Lucknow C	1278 kHz	MW Vividh Bharati service
11	Mumbai C	1188 kHz	MW Vividh Bharati service
12	Panaji B	1539 kHz	MW Vividh Bharati service
13	Vijayawada B	1503 kHz	MW Vividh Bharati service
14	Varanasi B	1602 kHz	MW Vividh Bharati service
15	Ahmedabad	96.7 MHz	FM Vividh Bharati service
16	Allahabad	100.3 MHz	FM Vividh Bharati service
17	Bangalore	102.9 MHz	FM Vividh Bharati service
18	Bhopal	103.5 MHz	FM Vividh Bharati service
19	Chandigarh	103.1 MHz	FM Vividh Bharati service
20	Dharwad	103.0 MHz	FM Vividh Bharati service
21	Guwahati	100.8 MHz	FM Vividh Bharati service
22	Hyderabad	102.8 MHz	FM Vividh Bharati service
23	Indore	101.6 MHz	FM Vividh Bharati service
24	Jabalpur	102.9 MHz	FM Vividh Bharati service
25	Jaipur B	100.3 MHz	FM Vividh Bharati service
26	Jammu B	104.5 MHz	FM Vividh Bharati service
27	Jamshedpur	100.8 MHz	FM Vividh Bharati service
28	Jodhpur	102.1 MHz	FM Vividh Bharati service
29	Kochi B	107.5 MHz	FM Vividh Bharati service
30	Kozhikode	103.6 MHz	FM Vividh Bharati service
31	Nagpur	100.6 MHz	FM Vividh Bharati service
32	Patna	102.5 MHz	FM Vividh Bharati service
33	Pune	101.0 MHz	FM Vividh Bharati service
34	Rajkot	95.8 MHz	FM Vividh Bharati service
35	Ranchi	103.3 MHz	FM Vividh Bharati service
36	Siliguri	101.4 MHz	FM Vividh Bharati service
37	Srinagar	102.6 MHz	FM Vividh Bharati service
38	Surat	101.1 MHz	FM Vividh Bharati service
39	Thiruvananthapuram	101.9MHz	FM Vividh Bharati service
40	Vadodara	93.9 MHz	FM Vividh Bharati service
41	Visakhapatnam	102.0 MHz	FM Vividh Bharati service
42	Bangalore (Amrutha Varshini)	100.1 MHz	FM Vividh Bharati service

The AIR Vividh Bharati transmission on SW channel is at 10.33 MHz. Other services offered by AIR includes 24-hour, *free-to-air direct-to-home (DTH)* service, *radio-on-demand (ROD)* service on FM transmitters, and digital satellite radio service. [4]

Exercises

1. A 1 MHz carrier is simultaneously modulated with 500, 800 and 3 kHz audio sine wave signals. What will be the frequencies present at the output? (Answer: 1 MHz, 1 MHz $\pm$ 500 Hz, 1 MHz $\pm$ 800 Hz and 1 MHz $\pm$ 3000 Hz)
2. When the percentage modulation is 80, an AM transmitter generates 12 kW.
   (a) How much of this is the carrier power?
   (b) If the carrier and one of the sidebands were suppressed, what would be percentage of power saving? (Answer: 9.091 kW, 87.88%)
3. For a carrier power of 10 kW, plot the total transmitter power for modulation index varying from 0 to 100%. Use MATLAB for coding.
4. The output power of a 50% modulated AM generator is 1.8A.
   (a) To what value will this current rise if it is simultaneously modulated by another signal to the extent of 80%?
   (b) What will be the percentage saving in power if the carrier and one of the sidebands are suppressed before transmission? (Answer: 2.04 A, 84.60%)
5. A transistor Class C amplifier has the maximum permissible collector dissipation of 40W and a collector efficiency of 75%. It is to be collector modulated to the depth of 90%.
   (a) Calculate the maximum unmodulated carrier power,
   (b) The sideband power generated.
   (c) If the depth of modulation is now reduced to 70%, calculate the sideband power. (Answer: 30W, 12.15W & 7.35W)
6. Let $m(t)$ be a rectangular signal having an amplitude of 1 and a pulse-duration of 50 ms. The carrier signal frequency is 3 kHz. Assume that the phase of the carrier signal is zero and the amplitude of the carrier signal is 1.
   (a) Show the spectrum of the message signal, $M(f)$.
   (b) Show the spectrum of the DSB-SC AM modulated signal, $U(f)$.
7. Let $m(t) = \sin c(200t)$ for $|t| < 50$ ms, and = 0 for otherwise. Assume that the phase of carrier signal is zero and the amplitude of the carrier signal is 1, and the carrier signal frequency is 1 kHz.
   (a) Show the spectrum of the message signal, $M(f)$.
   (b) Show the DSB-SC AM modulated signal both in time domain and in frequency domain.

[4] Source: *www.allindiaradio.org*

8. Let $m(t) = \sin(2\pi f_m t)$, for $0 \le t \le 10$ ms and 0 for otherwise, where f_m is 1 kHz. The carrier frequency is 10 kHz. Assume that the phase of carrier signal is zero and the amplitude of the carrier signal is 1.
   (a) Show the spectrum of the message signal, $M(f)$.
   (b) Show the DSB-SC AM modulated signal both in time domain and in frequency domain.
9. Let $m(t) = 2\sin(2\pi f_m t)$, for $0 \le t \le 10$ ms and 0 for otherwise, where f_m is 200 Hz. The carrier frequency is 3 kHz. Assume that the phase of carrier signal is zero and the amplitude of the carrier signal is 1.
   (a) Show the spectrum of the message signal, $M(f)$.
   (b) Show the USB-AM modulated signal both in time domain and in frequency domain.
   (c) Show the LSB-AM modulated signal both in time domain and in frequency domain.
10. Let $m(t) = \frac{1}{2}\sin(2\pi f_m t)$, for $0 \le t \le 10$ ms and 0 for otherwise, where f_m is 400 Hz. The carrier frequency is 5 kHz. Assume that the phase of carrier signal is zero and the amplitude of the carrier signal is 1.
    (a) Find the modulation index, m_a, mathematically.
    (b) Show the conventional AM modulated signal in both time domain and frequency domain.
11. Let $m(t) = A\sin(2\pi f_m t)$, for $0 \le t \le 10$ ms and 0 for otherwise, where f_m is 200 Hz. The carrier frequency is 10 kHz. Assume that the phase of carrier signal is zero and the amplitude of the carrier signal is 1.
    (a) Find the modulation index, m_a, mathematically.
    (b) When A is equal to 0.5, show the modulated signal in time domain with the modulation index used.
    (c) When A is equal to 2, show the modulated signal in time domain with the modulation index used.
    (d) Explain the effect of overmodulation.
12. A lowpass signal $x(t)$ with a bandwidth of 10 kHz is modulated with a carrier $\cos\omega_c t$ to produce $s(t)$. Find the value of f_c so that the bandwidth of $s(t)$ is 1% of f_c. (Answer: 2 MHz)
13. The information signal $x(t) = 2\sin 2000\pi t + \cos 4000\pi t$ is modulated by a DSB modulator operating at a carrier frequency of 100 kHz. Sketch the power spectral density of the output. Verify your answer by using MATLAB tools. (Answer: 4@100 kHz ± 1 kHz, 1@100 kHz ± 2 kHz)
14. Consider the block diagram shown in Figure 3.16 and a triangular pulse train with a period of 1.0 μs.
    (a) Show that this scheme will work if the local carrier $g(t)$ has no dc component. The filter cut-off frequency is $f_c + f_x$, with f_c being the carrier frequency and f_x the bandwidth of the input.

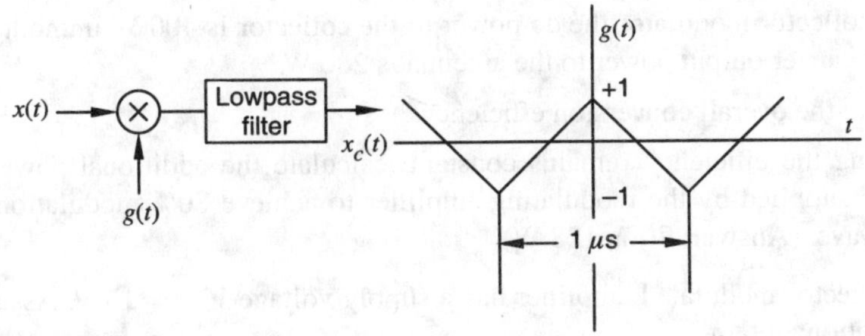

Figure 3.16 *A simple AM modulator*

(b) If the input is a single tone sinusoid $x(t) = 2\cos 1000\pi t$, find the required filter bandwidth.

(c) Write an expression for the output of the filter.

(d) How would you modify the system if the local carrier $g(t)$ had a dc bias of 2.0 volts?

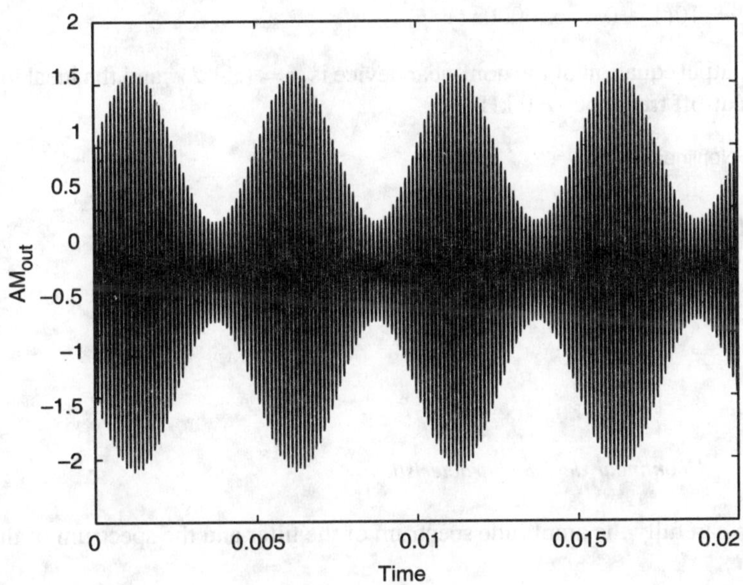

Figure 3.17 *An AM DSB signal*

15. A sinusoidally modulated DSB AM waveform is shown in Figure 3.17.

(a) Find its modulation index.

(b) Compute the sideband power percentage η.

(c) Determine the amplitude of the carrier to attain a modulation index of 0.1. (Answer: 33.33%, 5.56%, 25V)

16. In a class C collector modulator the dc power to the collector is 400 W unmodulated, for which the carrier output power to the antenna is 280 W.
    (a) Calculate the overall conversion efficiency.
    (b) Assuming the efficiency remains constant, calculate the additional power that must be supplied by the modulating amplifier to achieve 80% modulation with a sine wave. (Answer: 70%, 128 W)

17. A class C collector modulated amplifier has a supply voltage $V_{cc} = 15$ V. Assuming 100% modulation, obtain
    (a) the minimum voltage rating of the transistor,
    (b) if the load resistance R_L is $50\Omega s$ and the conversion efficiency, η is 75%, the output power at 100% modulation. (Answer: 60 V, 2.53 W)

18. A modulating signal $e_m = \sin\omega_m t + 4\sin 4\omega_m t$ is used to multiply a carrier signal given by $e_c = 10\sin\omega_c t$, where all voltages are in volts and $f_c = 500$ kHz and $f_m = 2$ kHz. The multiplier constant is $k = 1\ \text{V}^{-1}$. Draw the spectrum of the modulated wave and determine the average power if $R_L = 50\,\Omega$.

19. The input to a square-law device of the system illustrated in Figure 3.18 is given by

$$x_c(t) = 10(1 + 0.4\cos 6000\pi t)$$

The input–output equation of the nonlinear device is $y = (x+2)^2$ and the ideal lowpass filter has a cut-off frequency 8.0 kHz.

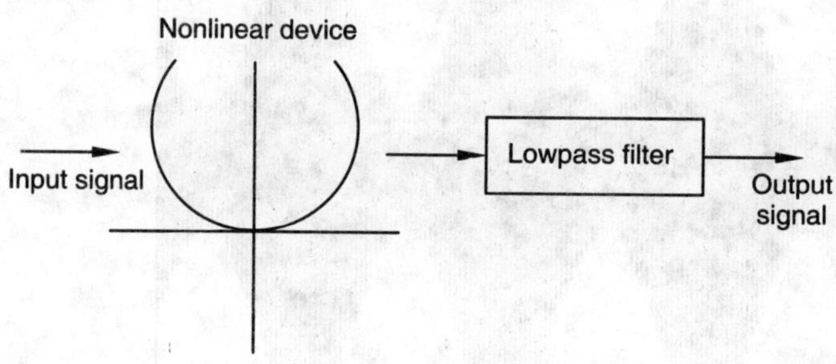

Figure 3.18 *A nonlinear device characteristics*

    (a) Sketch carefully the amplitude spectrum of the filter and the spectrum of the output $y(t)$.
    (b) Verify your answers using MATLAB.

20. Show that the following system shown in Figure 3.19 can demodulate an AM signal of the form: $x_{AM}(t) = (A + m(t))\cos\omega_c t$ regardless of the value of A.

21. A sinusoidal carrier is amplitude modulated by a square wave that has zero dc component, and a peak-to-peak value of 2 V. The period of the square wave is 0.5 ms. The carrier amplitude is 2.5 V and its frequency is 10 kHz.
    (a) Write the equations for the modulating signal and the modulated signal.
    (b) Plot the above signals over time and obtain the modulation index.

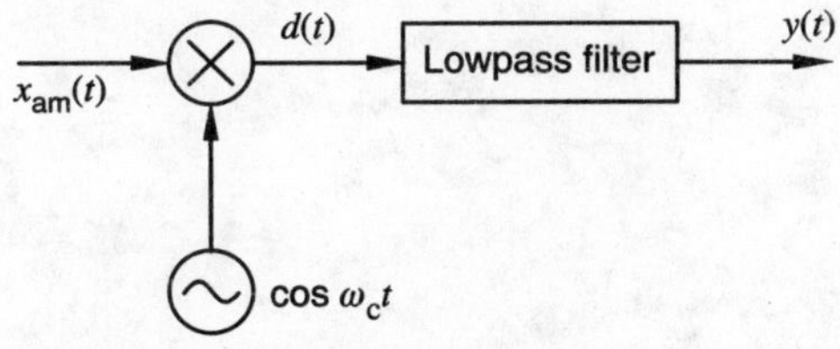

Figure 3.19 *A generic AM demodulator*

22. Two sine waves simultaneously amplitude-modulate a carrier. Signal 1 is of amplitude 3 V and frequency 300 Hz, while the second one is of amplitude 4 V and frequency 800 Hz. The carrier amplitude is 10 V and its frequency is 20 kHz. Plot the modulating waveform and the AM_{out} signal using MATLAB.

Chapter 4

AM Transmitters and Receivers

The basic principles of amplitude modulation were introduced in the previous chapter. However, we did not attempt a detailed study of the transmitters and receivers used in amplitude modulation (AM) and demodulation. This chapter is devoted to such a study. We begin with a discussion on practical AM broadcast transmitters which occupy the band of frequencies from 535 kHz to 1605 kHz in medium wave (MW) band vis-à-vis their design principles and typical circuits. It will be followed by the principle of superheterodyning and various AM demodulators. The single sideband techniques are also discussed. More recent trends in broadcasting technology like the digital amplitude modulation, Digital Radio Mondale, and the electronically tuned receivers are covered. The chapter ends with several modelling examples using MATLAB.

4.1 Transmitters

Amplitude modulators can generally be classified into two: low-level and high-level modulators. The block diagrams of both are given in Figures 4.1 and 4.2, respectively. In practical circuits, if the modulating signal is applied to base or emitter (gate or source in the case of an FET; gate or cathode in the case of vacuum tubes), low-level modulation results. On the other hand if the modulating voltage is applied to collector (drain in the case of an FET or plate in the case of vacuum tube), high-level modulation results.

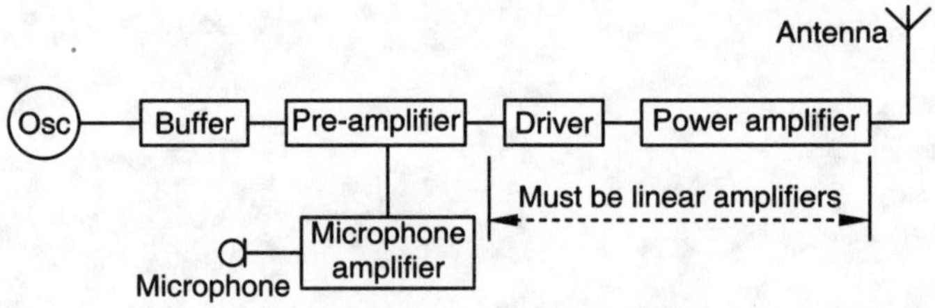

Figure 4.1 *Block diagram of low-level modulator*

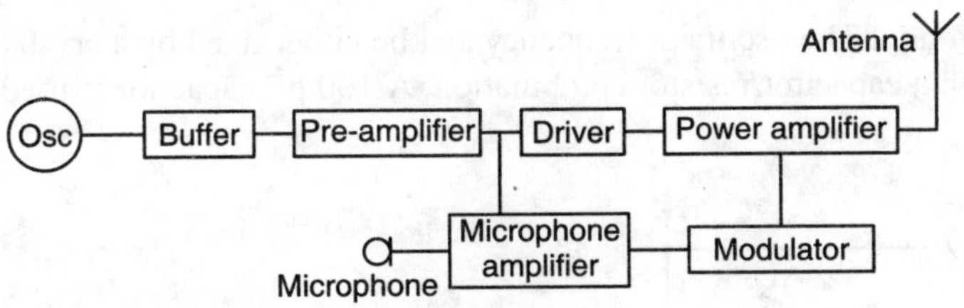

Figure 4.2 *Block diagram of high-level modulator*

4.1.1 Low- and high-level modulators

Note that in low-level modulators, all stages after the modulator must be linear. The advantage of low-level AM is that *not much audio power is required.* The disadvantage is that all radio frequency (RF) amplifiers after the modulator (pre-amplifier) *must be linear*, and linear amplifiers are low-efficiency class A or class B push-pull power amplifiers. Low-level AM is best suited for low-power transmitters such as handhelds. Now there are ICs available that contain the whole modulator circuit on a single chip.[1] A balanced-modulator-based low-level AM generator is given in Figure 4.3. In high-level modulation, the modulation occurs in the

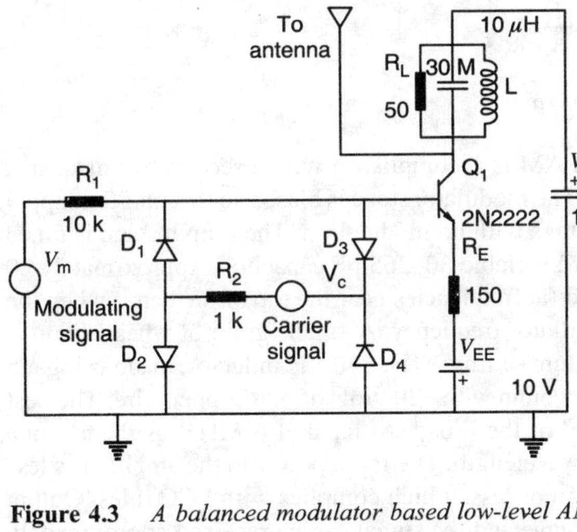

Figure 4.3 *A balanced modulator based low-level AM generator*

power amplifier stage which is always class C type. The advantage is that the carrier signal from the oscillator can be amplified using the high-efficiency class C amplifier. The disadvantage is that a large amount of audio amplification is required to modulate the carrier at high-power. High-level modulators are invariably used in high-power AM transmitters.

4.1.2 Micropower AM transmitter

A micropower AM transmitter circuit is given in Figure 4.4. In this circuit, a 74HC14 hex Schmitt trigger inverter is used as a square wave oscillator to drive a small signal transistor

[1] For example, the AM-RT4-315 IC from RF Solutions Ltd.

in a class C amplifier configuration. The oscillator frequency can be either fixed by a crystal or made adjustable (VFO) with a capacitor/resistor combination. A 100 pF capacitor is used

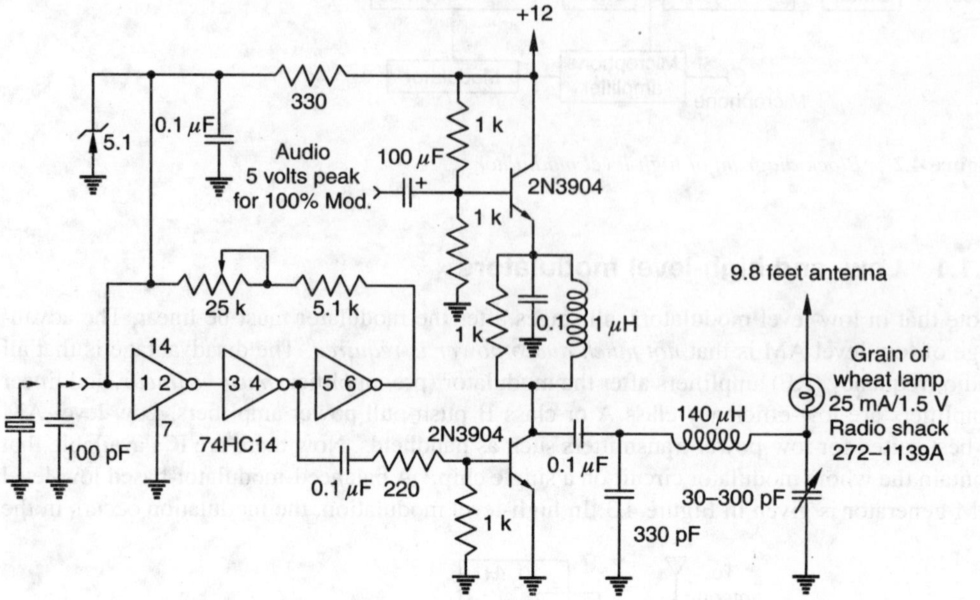

Figure 4.4 *A micropower low-level AM generator*

in place of the crystal for VFO operation. AM is accomplished with a second transistor that controls the dc voltage to the output stage. The modulator stage is biased so that half the supply voltage or 6.0 V is applied to the output stage with no modulation. The output stage is tuned and matched to the antenna with a standard variable 30–365 pF capacitor. Approximately 20 mA of current will flow in the antenna lead (at frequencies near the top of the band) when the output stage is optimally tuned to the oscillator frequency. A small 'grain of wheat' lamp is used to indicate antenna current and optimum settings. The 140 μH inductor, made using a 5 cm length of 26.67 cm (OD) PVC pipe, is wound with 120 turns of #28 copper wire. The best performance is obtained near the high end of the broadcast band (1.6 MHz) as the antenna length is only a very small fraction of the wavelength. The input power to the amplifier is less than 100 mW and the antenna length is 3 m or less, which complies with FCC rules. Output power is somewhere in the 40 microwatt range and the signal can be received approximately within a radius of 24.38 m. Radiated power output can be approximated by working out the antenna radiation resistance and multiplying it by the antenna current squared. The radiation resistance for a dipole antenna with length less than $\frac{1}{4}$th wavelength is:

$$R = 80\pi^2 \times \left[\frac{\text{Length}}{\text{Wavelength}}\right]^2 \times \left[\frac{\text{average current along the rod}}{\text{feed current}}\right]^2 \tag{4.1.1}$$

Even if the rods are capped with plates, this factor cannot be larger than 1. Substituting values for a 2.987 m dipole at a frequency of 1.6 MHz, we get $R = 790 \times 0.000354 \times 0.25 = 0.07\ \Omega$. And the resistance will be only half as much for a monopole or $0.035\ \Omega$. Radiated power at 20 mA works out to about $I^2R = 14$ mW.

4.1.3 AM based on analog multipliers

An integrated solution to a low-cost IC that can multiply two analog signals is the AD534 IC from Analog Devices, Inc. A practical modulator circuit using it has been depicted in Figure 4.5.[2] Another implementation makes use of logarithmic amplifiers, a voltage summer, and an exponential circuit to multiply signals to generate AM. This can be considered as an indirect analog multiplier method. As a note we may add that in most of the practical systems the modulation is carried out in more than one stages—first the baseband audio modulates a *sub-carrier signal*, which will in turn modulate the final carrier. In the receiver the demodulation process is carried out in the reverse order.

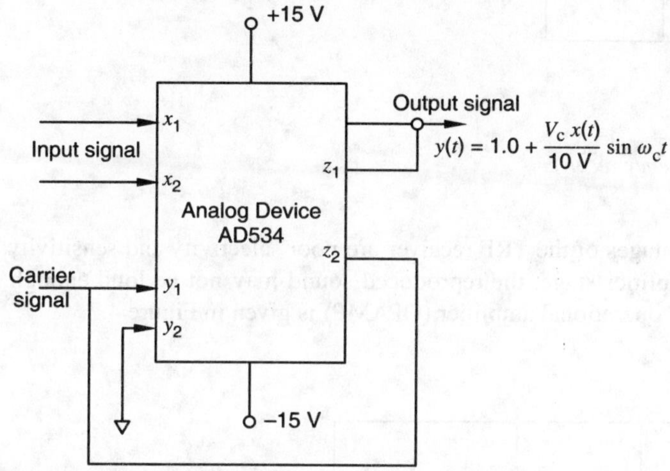

Figure 4.5 *A modulator based on AD534 analog multiplier*

4.1.4 Antenna matching unit

The power from the final stage of power amplifiers is fed to the antenna matching unit (AMU). The purpose of the AMU is to match the impedance of the radiating element to that of the power amplifier. Unless the impedances are properly matched at every stage, RF energy will be reflected from the unbalanced point which can even damage the previous stage. It may also be noted that the antenna impedance has to be matched, in turn, to the *intrinsic impedance of free space*, which is 377 Ω. For commercial AM broadcast transmission, a *quarterwave dipole* antenna, known as *Marconi antenna*, is used.

4.2 Simple Tuned Radio Frequency Receivers

The tuned radio frequency (TRF) receivers use a simple LC tuneable circuit, along with a coil antenna, a detector diode (say, OA79), and a headphone/loud speaker. It is also popularly known as a 'crystal receiver.' The parallel tuned LC circuit resonates only at the frequencies of the stations, thus selecting them and rejecting others. A typical crystal receiver is illustrated in

[2]Courtesy Analog Devices Inc.

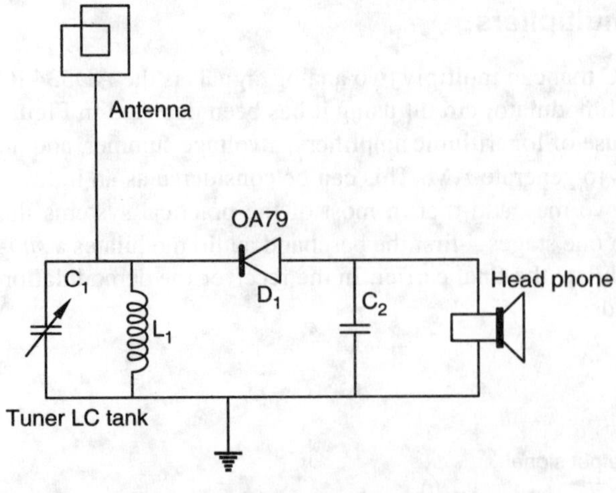

Figure 4.6 *A tuned radio frequency receiver*

Figure 4.6. The major disadvantages of the TRF receiver are poor selectivity and sensitivity. Also, as there is no power amplifier stage, the reproduced sound may not be loud enough. Another TRF receiver using an operational amplifier (OPAMP) is given in Figure 4.7.

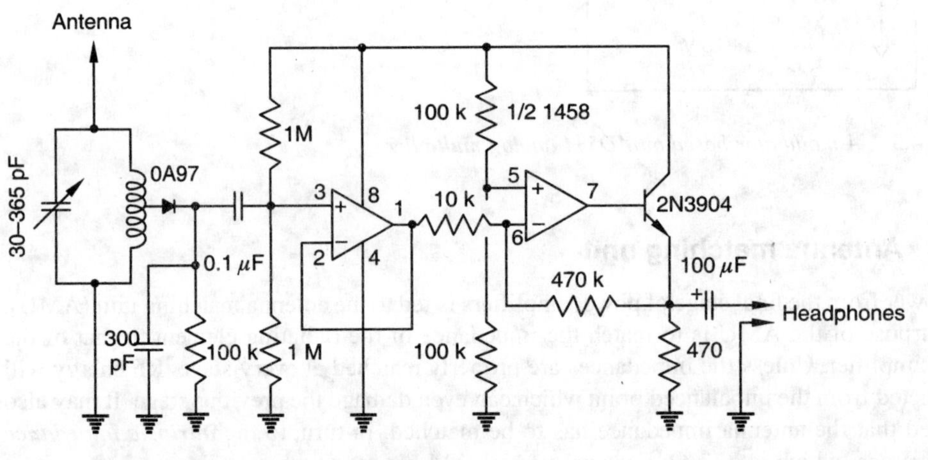

Figure 4.7 *A TRF receiver using OPAMP*

This is basically a crystal radio with an audio amplifier that is fairly sensitive and receives several strong stations, with a minimal 4.572 m antenna. Longer antennas will provide stronger signals but the selectivity will not be good and strong stations may be heard in the background of weaker ones. Using a long-wire antenna, the selectivity can be improved by connecting it to one of the taps on the coil instead of the junction of the capacitor and coil. Some connection to ground is required. The inductor in this case was wound with 200 turns of #28 enamelled copper wire on a 2.2 cm diameter, 10.16 cm length of PVC pipe, which yielded about 220 μH. The inductor was wound with taps every 20 turns so that the diode and antenna connections could be selected for best results—this happened to be 60 turns from the antenna-end for the diode. The diode should be a germanium (1N34A type) one for best results, but silicon diodes will also work if the signal is strong enough. The carrier frequency is removed from the rectified signal at the cathode of the diode by the 300 pF capacitor; the audio frequency is passed by the 0.1 μF capacitor to the non-inverting input of the first OPAMP, which functions as a high-impedance buffer stage. The second OPAMP stage increases the voltage level about 50 times and is dc coupled to the first through the 10k resistor. If the pairs of 100 kΩ and 1 MΩ resistors are not close in value (1%), one may need to either use closer matched values or add a capacitor in series with the 10k resistor to keep the dc voltage at the transistor emitter between 3 and 6 V. Another approach would be to reduce the overall gain with a smaller feedback resistor (470 k Ω). High-impedance headphones will probably work best, but walkman stereo type headphones will also work. The circuit draws about 10 mA current from a 9 V source.

The problems of instability, insufficient adjacent frequency rejection, and bandwidth variation can all be solved by the use of a superheterodyne receiver, which introduces relatively few problems of its own.

4.3 Superheterodyne Receivers

In practice, the commercial AM receivers have a two-stage reception structure. The first stage is called the RF stage and is tuneable. The second stage, intermediate frequency IF stage, is a fixed design common to all receivers. *In a practical system, the receiver has to perform the task of demodulation while achieving the following:*

1. Carrier frequency tuning: the selection of the desired RF signal (in radio or television)
2. Filtering: the separation of desired signal from other modulated signals that are booked along with the desired signal
3. Amplification: to compensate for the loss of signal power incurred from transmitted point to the received point

This architecture is known as the superheterodyne receiver. Figure 4.8 illustrates this structure.

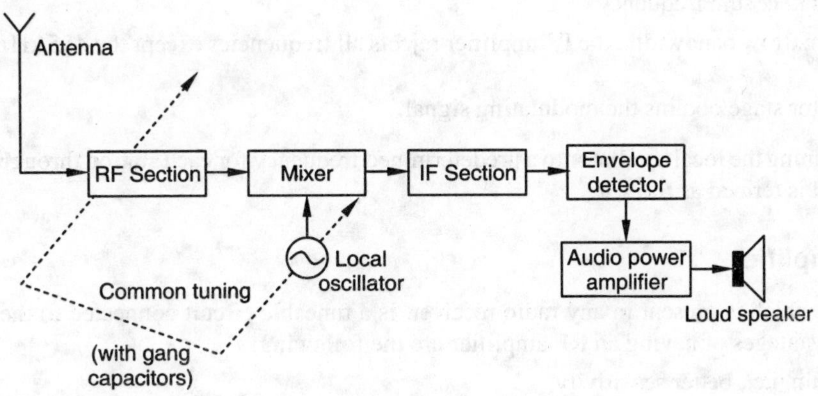

Figure 4.8 *Block diagram of superheterodyne receiver*

The local oscillator frequency, $f_{\rm LO}$ is the sum of the carrier frequency ($f_{\rm c}$) and the IF frequency, $f_{\rm IF}$ (which is normally a fixed frequency). For the AM broadcast in MW band, the IF frequency is, for example, 455 kHz. So, we have

$$f_{\rm LO} = f_{\rm c} + f_{\rm IF} = f_{\rm c} + 455 \text{ kHz} \qquad (4.3.2)$$

A constant frequency difference is maintained between the local oscillator and the RF circuits, in which all the capacitors are ganged together and operated in unison by one control knob. The IF amplifier provides the gain and hence the sensitivity and bandwidth requirements of the receiver.

With reference to Figure 4.8, the RF stage is a wideband RF amplifier tuneable from approximately 530 kHz to 1.62 MHz (the standard commercial MW band). It is mechanically tied to the local oscillator to ensure precise tuning characteristics.

The local oscillator is a variable oscillator capable of generating a signal from 0.985 MHz to 2.075 MHz. The incoming signal from the transmitter is selected and amplified by the RF stage. It is then mixed with a predetermined local oscillator signal in the mixer stage. A class C nonlinear device produces the sum, difference, and original signals as explained elsewhere.

The output of the mixer is fed to the IF amplifier. This is a very narrow bandwidth class A (linear) amplifier capable of selecting a frequency of 0.455 MHz $\pm$ 3 kHz and rejecting all others.

The next stage is the detector, which eliminates one of the sidebands still present and separates the RF from the audio components of the other sideband. The RF is filtered to ground and the audio fed to audio amplifier stages and subsequently to the loudspeakers. The various steps involved in receiving a station working at a frequency of 700 kHz is illustrated below:

- Tune the RF amplifier to the lower end of the MW band.
- Tuning the RF amplifier to 700 kHz also tunes the local oscillator to a predetermined frequency of 700 kHz + 455 kHz = 1155 kHz.
- The two signals are mixed in the mixer stage, which produces the following signals at the output of the mixer:
    1. 1155 kHz local oscillator frequency
    2. 700 kHz AM station carrier frequency
    3. 455 kHz difference frequency
    4. 1855 kHz sum frequency
- Due to its narrow bandwidth, the IF amplifier rejects all frequencies except the 455 kHz signal.
- The detector stage obtains the modulating signal.

The process of tuning the local oscillator to a predetermined frequency for each station throughout the AM band is termed as *tracking*.

4.3.1 RF amplifier

The RF section, which is present in any radio receiver, is a tuneable circuit connected to the antenna. The advantages of having an RF amplifier are the following:

1. Greater gain, i.e., better sensitivity
2. Improved image frequency rejection
3. Improved signal-to-noise ratio

4. Improved rejection of unwanted adjacent channels, i.e., better selectivity
5. Better coupling of receiver to antenna
6. Prevention of spurious frequencies from entering the mixer and heterodyning to produce an interference frequency equal to the IF from the desired signal
7. Prevention of re-radiation of the local oscillator through the antenna of the receiver (relatively rare)

Single tuned transformer-coupled amplifiers are commonly used in RF amplifiers.

Sensitivity

The sensitivity of a radio receiver is its ability to amplify weak signals. It is often defined in terms of the voltage that must be applied to the receiver input terminals to give standard output power, measured at the output terminals. For AM broadcast, an audio frequency of 400 Hz with a modulation percentage of 30% is used, and the signal is applied to the receiver through a standard coupling network known as *dummy antenna.* The standard output is 17 dB μ, and for all types of receivers the loud speaker is replaced by a load resistance of equal value (say, 8 Ω). The sensitivity of a receiver is expressed in microvolts or decibels below 1V and measured at three points along the tuning range during production. A typical value can be 16.4 μV or –95.7 dBV.[3] The most important factors determining the sensitivity of a superheterodyne receiver are the gain of the IF and RF amplifier stages.

Selectivity

The selectivity of a receiver is its ability to reject unwanted signals. The bandpass response nature of the RF and IF stages decides the selectivity.

4.3.2 Rejection of image frequency

In all standard broadcast AM receivers, the local oscillator frequency is $f_{\mathrm{LO}} = f_{\mathrm{s}} + f_{\mathrm{IF}}$, or $f_{\mathrm{s}} = f_{\mathrm{LO}} - f_{\mathrm{IF}}$. When f_{s} and f_{LO} are mixed, the difference frequency, which is one of the by-products, is equal to f_{IF}. As such, it is the only frequency that is passed and amplified by the IF stage.

If a frequency f_{si} reaches the mixer such that $f_{\mathrm{si}} = f_{\mathrm{LO}} + f_{\mathrm{IF}}$, that is $f_{\mathrm{si}} = f_{\mathrm{s}} + 2f_{\mathrm{IF}}$, then this frequency will also produce f_{IF} when mixed with f_{LO}. Unfortunately, this spurious frequency will also be amplified by the IF amplifier and will result in interference. This has the effect of receiving two stations simultaneously and is naturally undesirable. The frequency f_{si} is called the *image frequency*.

The rejection of the image frequency by a single tuned circuit, i.e., the ratio of the gain at the signal frequency to the gain at the image frequency is given by

$$\alpha = \sqrt{1 + Q^2\rho^2} \tag{4.3.3}$$

where

$$\rho = \frac{f_{\mathrm{si}}}{f_{\mathrm{s}}} - \frac{f_{\mathrm{s}}}{f_{\mathrm{si}}}. \tag{4.3.4}$$

[3]The sensitivity in dBV is calculated as Sens-in-dBV $= 20\log_{10}\left(\frac{\text{Voltage}}{\text{1volt}}\right)$.

Q = the loaded Q of the tuned circuit.

If the receiver has an RF stage, then there are two tuned circuits, both tuned to f_s. The rejection of each can be calculated by the preceding equation and the total rejection will be the product of the two. *Image rejection* depends on the front-end selectivity of the receiver and must be achieved before the IF stage.

Example 4.1 An AM broadcast receiver has an IF of 455 kHz and is tuned to 1200 kHz and the RF stage has a tuned circuit with a Q of 80.

1. Obtain the image frequency.
2. Find the image rejection in decibels.

Solution

For part (a):

For

$$f_s = 1200\text{kHz},$$

$$f_{LO} = f_s + 455\text{kHz} = 1200 + 455 = 1655\,\text{kHz}$$

and

$$f_{si} = f_{LO} + f_{IF} = 1655 + 455 = 2110\,\text{kHz}.$$

For part (2):

$$\rho = \frac{f_{si}}{f_s} - \frac{f_s}{f_{si}},$$

$$= \frac{2110}{1200} - \frac{1200}{2100} = 1.1896.$$

$$\alpha = \sqrt{1 + Q^2\rho^2} = \sqrt{1 + 80^2 \times 1.1896^2} = 95.17 = 39.57dB$$

Example 4.2 Obtain the image rejection of the above receiver when the signal frequency is 25 MHz. In order to make the image frequency rejection the same as in Example 4.1, calculate the loaded Q that the RF amplifier must have.

Solution

For part (1):

$$\rho = \frac{25 + 2 \times 0.455}{25} - \frac{25}{25 + 2 \times 0.455} = 0.0715$$

$$\alpha = \sqrt{1 + 80^2 \times 0.0715^2} = 5.81 = 15.28dB$$

For part (2):

$$\alpha = 95.17 = \sqrt{1 + Q_L^2 \times 0.0715^2}$$

$$\Rightarrow Q_L^2 = \frac{95.17^2 - 1}{0.0715^2}$$

$$\Rightarrow Q_L = \sqrt{1770586.7} = 1330.634.$$

Double spotting

This is a well-observed phenomenon that manifests itself by picking up of the same shortwave station at two nearby points on the receiver dial. It is caused by inadequate image frequency rejection. It is assumed that there is no RF section in receivers badly suffering from double spotting. Consider such a receiver having an IF of 455 kHz. If there is a strong station at 10 MHz, then the receiver will naturally pick it up, with the local oscillator frequency being set to 10.455 MHz. The receiver will also pick up the strong station when the receiver is tuned to 9.09 MHz. When the receiver is tuned to 9.09 MHz, its local oscillator will be tuned to 9.545 MHz. As this is exactly 455 kHz below the frequency of the strong station, the two signals will produce 455 kHz when they are mixed; the IF amplifier will not reject this signal. Had there been an RF amplifier, the 10 MHz signal might have been rejected before reaching the mixer.

Double spotting may be used to calculate the IF frequency of an unknown receiver since the spurious point on the dial is precisely $2f_{\text{IF}}$ below the correct frequency. An improvement in image frequency rejection will produce a corresponding reduction in double spotting.

4.3.3 Local oscillator

In receivers operating up to the limit of shortwave broadcasting, that is 36 MHz, the most common types of local oscillators are the *Armstrong* and the *Hartley*. The *Colpitts*, *Clapp*, or the ultra-audio oscillators are used at the top of this range and above. When the frequency stability of the local oscillator is important, automatic gain control (AGC) or frequency synthesisers are used.

4.3.4 IF amplifiers

The following factors decide the choice of IF in any system:

- If the IF is too high, poor selectivity and poor adjacent channel rejection results unless sharp cutoff filters are used in the IF stages.
- A high value of IF increases tracking problem.
- As the IF is lowered, image frequency rejection becomes poor. It is seen that the image frequency rejection worsens as the signal frequency is raised.
- A very low IF can make the selectivity too sharp, cutting off sidebands.
- If the IF is too low, the frequency stability of the local oscillator must be made correspondingly higher because any frequency drift is now a larger proportion of the low IF than a high IF.

- The IF must not fall within the tuning range of the receiver because instability will occur and heterodyne whistles will be heard, making it impossible to tune to the frequency band immediately adjacent to the IF.

Frequencies used

IF values for typical applications based on many years' experience, are given below:

1. Standard MW AM broadcast: IF in the range 438–465 kHz, 455 kHz being the most popular.
2. Shortwave AM (single side-band (SSB) or VHF reception): First IF in the range 1.6–2.3 MHz or above 30 MHz.
3. Frequency modulation (FM) receivers (with a band of 88–108 MHz): IF fixed at 10.7 MHz.
4. Television receivers in the VHF band (54–223 MHz) and UHF band (470–940 MHz): IF between 26 and 46 MHz, 36 MHz and 46 MHz being two popular values.
5. Microwave and RADAR receivers operating in the 1–10 GHz range: IF which is either 30 MHz, 60 MHz or 70 MHz.

IF amplifier stage

The IF amplifier is a fixed frequency amplifier with the very important function of rejecting adjacent unwanted frequencies. It should have frequency response with steep edges. The IF amplifier can be either single tuned or double tuned.

4.3.5 Automatic gain control

When a receiver without automatic gain control (AGC) is tuned to a strong station, the signal may overload the IF and audio frequency stages, causing severe distortion and a disturbing blast of sound. The AGC derives a varying bias signal that is proportional to the average received signal and uses this bias to vary the gain of one or more RF and IF stages. When the average level increases, the size of the AGC bias increases, and the gain of the controlled stages decreases. On the other hand, when there is no signal, there is a minimum AGC bias, and the amplifiers provide maximum gain.

A simple AGC is used in most domestic receivers and inexpensive communication receivers. In this, the AGC bias starts to increase as soon as the received signal level exceeds the background noise level, and the receiver immediately becomes less sensitive. The detector used in these cases will be a simple half wave rectifier that produces a dc level that is proportional to the average signal level. This dc level is passed through an RC lowpass filter circuit to remove the audio signal and is then applied to the base of the IF and RF stages for biasing. The time constant of the filter must be at least 10 times larger than the period of the lowest modulating frequency received, which is usually 50 Hz. A typical simple AGC circuit is shown in Figure 4.9.

Delayed AGC

A better response can be obtained by including more gain in the feedback loop. This is done by including a dc amplifier stage after the AGC filter. Delayed AGC is used in most of the better category communication receivers. In systems employing delayed AGC, the generation

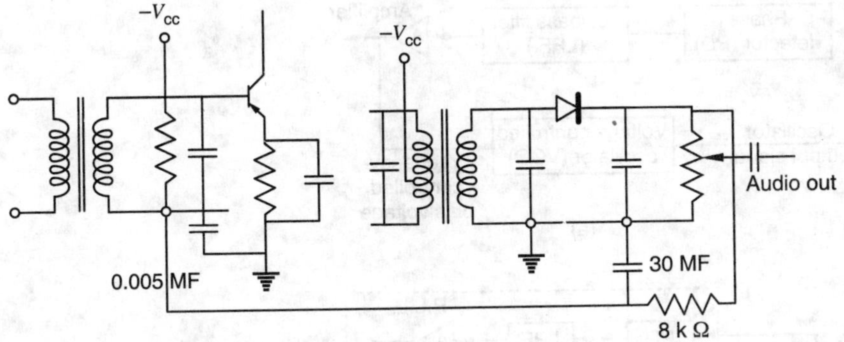

Figure 4.9 *A typical AGC circuit*

of AGC bias is prevented until the signal level exceeds a preset threshold and increases proportionally there on. This results in maximum sensitivity of the receiver for low-level signals.

4.3.6 Double conversion receivers

The front end of any AM receiver must reject the first image frequency located at 2IF from the desired signal frequency. For this, the IF value must be high. However, as the IF is made higher, its bandpass response becomes wider (because of the changing Q of the tuned circuits). Beyond the high-frequency band (30 MHz) it becomes impossible to obtain the required bandpass response and image rejection using ordinary tuned circuits. Therefore, single conversion superheterodyne receivers are seldom used above 20 MHz.

The *double-conversion receiver* allows the receiver to have good image rejection and also good adjacent channel selectivity. The image rejection is obtained in the first conversion stage by assuring that the image frequency is well outside the fixed-tuned RF amplifier bandpass limits. The narrow bandpass response required for good adjacent channel rejection is obtained in the second IF, which may as low as 100 kHz. It should be noted that the *high intermediate frequency* must come first. If this is not the case, the image frequency will be insufficiently rejected at the input and will be mixed up with the proper signal—no amount of high IF stages will make any difference afterwards.

4.3.7 AM receiver using phase-locked loop

The *phase-locked-loop* based AM demodulator is shown in Figure 4.10, where the block diagram of the phase-locked loop (PLL) is also shown. The operation of the PLL is quite simple. When the free running frequency of the *voltage controlled oscillator (VCO)* (VCO) is compared with the input signal at the phase detector, and if the two are very close, the PLL attains a lock. Now, for variation of the input signal within limits, the VCO output follows suit.

When the PLL locks onto the AM signal, the VCO frequency adjusts automatically to the AM carrier frequency. If the free-running frequency of the PLL (i.e. its frequency before lock-in) is close to the AM carrier frequency, the VCO output voltage is approximately 90° out of phase with the AM carrier wave. To compensate for the 90° phase difference that occurs within the PLL, the carrier is externally shifted by a further 90°, as shown in Figure 4.10. It does not matter whether the total phase difference is zero or 180°; the fundamental component

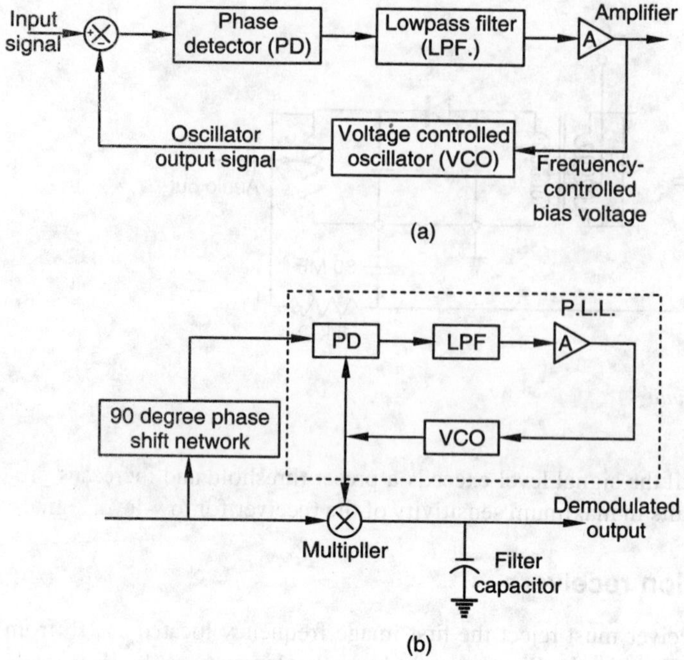

Figure 4.10 *The synchronous AM detector using phase-locked loop*

of the VCO is proportional to $\sin \omega_c t$, where ω_c is the angular frequency of the carrier.

Let the modulating signal be represented by *m(t)* and the AM signal therefore, by $(1 + m(t)) \cos \omega_c t \cos \omega_c t$. The output from the multiplier is proportional to

$$(1 + m(t))\cos \omega_c t \cos \omega_c t = \frac{(1 + m(t))}{2}(1 + \cos 2\omega_c t)\,. \tag{4.3.5}$$

The $\cos 2\omega_c t$ term is filtered out and the modulating signal is recovered from the term $\frac{m(t)}{2}$. This type of AM detection is known as *synchronous detection*.

4.4 Single Sideband Modulation

In AM, a carrier and two sidebands are produced. But even if the carrier and one of the side-bands are not transmitted, it is possible to retrieve the transmitted message at the receiver. The resulting signal will require less transmitted power and will occupy less bandwidth. This modulation technique is referred to as single sideband (SSB) transmission.

SSB modulation has been the fastest spreading form of analog communication in the second half of this century. Among its great advantages is the ability to transmit good communications quality signals by using a very narrow bandwidth with relatively low power.

4.4.1 SSB generation

The generation of the SSB signal involve two steps: the removal of the carrier and the removal of one of the side-bands. A balanced modulator is invariably used to remove the carrier. Diode-

or FET-based balanced modulator circuits are fairly popular. A typical circuit is given in Figure 4.11. A mathematical analysis of the functioning of the balanced modulator is as follows.

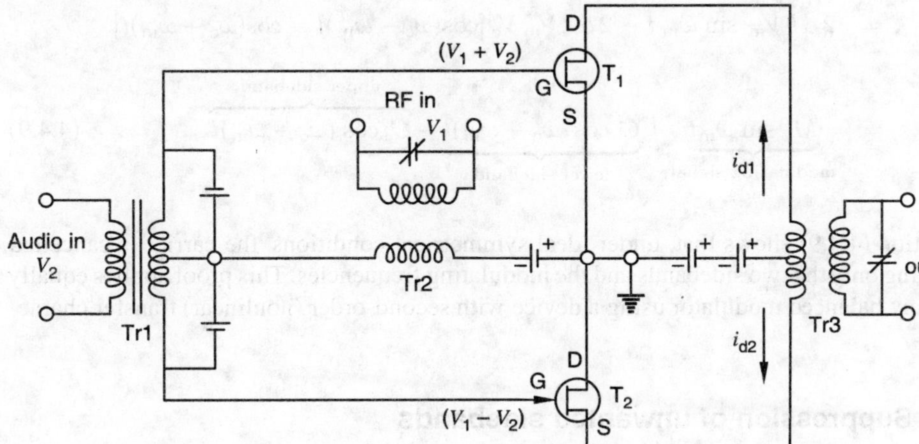

Figure 4.11 *A FET-based balanced modulator*

Note that the FET is a nonlinear device and if the gate voltage is v_g, then the drain current is given by $i_{d_1} = a + bv_g + cv_g^2$. Therefore, we have

$$\begin{aligned} i_{d_1} &= a + b(v_1 + v_2) + c(v_1 + v_2)^2 \\ &= a + bv_1 + bv_2 + cv_1^2 + cv_2^2 + 2cv_1v_2. \end{aligned} \tag{4.4.6}$$

$$\begin{aligned} i_{d_2} &= a + b(v_1 - v_2) + c(v_1 - v_2)^2 \\ &= a + bv_1 - bv_2 + cv_1^2 + cv_2^2 - 2cv_1v_2. \end{aligned} \tag{4.4.7}$$

Now the primary current of the transformer Tr3 is given by the difference between the individual drain currents. Thus, we have:

$$i_1 = i_{d_1} - i_{d_2} = 2bv_2 + 4cv_1v_2.$$

Now we can represent the carrier signal v_1 by $V_c \sin \omega_c t$ and the modulating signal v_2 by $V_m \sin \omega_m t$. Hence,

$$\begin{aligned} i_1 &= 2b\,V_m \sin \omega_m t + 4c\,V_m\,V_c\, \sin \omega_m t \sin \omega_c t \\ &= 2b\,V_m \sin \omega_m t + 4c\,V_m\,V_c\,\frac{1}{2}\left[\cos(\omega_c - \omega_m)t - \cos(\omega_c + \omega_m)t\right]. \end{aligned} \tag{4.4.8}$$

The output voltage v_0 is proportional to this primary current. Let the constant of proportionality be α. Then,

$$\begin{aligned} v_0 &= \alpha i_1 \\ &= 2\,\alpha\, b\, V_{\mathrm{m}}\,\sin\omega_{\mathrm{m}}\,t + 2\,\alpha\, c\, V_{\mathrm{m}}\, V_{\mathrm{c}}\,[\cos(\omega_{\mathrm{c}}-\omega_{\mathrm{m}})t - \cos(\omega_{\mathrm{c}}+\omega_{\mathrm{m}})t] \\ &= \underbrace{M\;\sin\omega_{\mathrm{m}}t}_{\text{modulating signal}} \;+\underbrace{C\,\cos(\omega_{\mathrm{c}}-\omega_{\mathrm{m}})t}_{\text{lower sideband}} - \overbrace{C\,\cos(\omega_{\mathrm{c}}+\omega_{\mathrm{m}})t}^{\text{upper sideband}} \end{aligned} \tag{4.4.9}$$

Equation (4.4.9) shows that, under ideal symmetrical conditions, the carrier is cancelled out, leaving only the two sidebands and the modulating frequencies. This proof applies equally well to any balanced modulator using a device with second-order (nonlinear) transfer characteristics.

4.4.2 Suppression of unwanted sidebands

All practical methods to generate SSB use the balanced modulator to suppress the carrier. But there are mainly three methods to remove the unwanted sideband. Let us study each one of them.

The filter method

This is the simplest of the three methods. After the balanced modulator, the unwanted sideband is removed (i.e., heavily attenuated) by a filter. The filter can be LC, crystal, or mechanical, depending on the carrier frequency and other constraints. A block diagram of an SSB transmitter employing filter method is given in Figure 4.12.

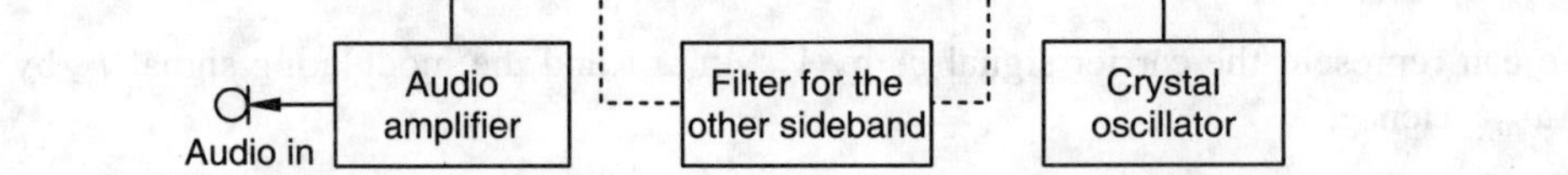

Figure 4.12 *Block diagram of an SSB transmitter using filters*

Out of the three types of filters, the mechanical filter seems to be the one with the best all-around properties of compact size, good bandpass, very good attenuation characteristics and an adequate upper frequency limit. Crystal and ceramic filters are comparatively inexpensive, but preferred only above 1 MHz.

The phase-shift method

In this method two balanced modulators and two phase-shifting networks are used. The block diagram of the system is given in Figure 4.13. Both the modulators produce an output consisting only of side-bands. The analysis of the system is as follows.

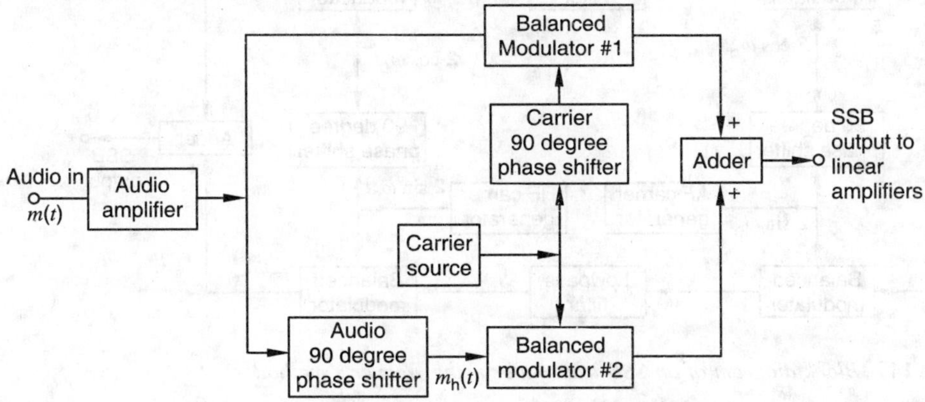

Figure 4.13 *Block diagram of an SSB transmitter using phase shifters*

The signal from the balanced modulator #1 is taken as v_1 and that from the second is taken as v_2. Let the audio signal be represented as $\sin \omega_m t$ and the carrier signal by $2 \sin \omega_c t$. Now,

$$v_1 = \cos[(\omega_c t + 90°) - \omega_m t] - \cos[(\omega_c t + 90°) + \omega_m t]$$

$$= \underbrace{\cos(\omega_c t - \omega_m t + 90°)}_{\text{LSB}} - \overbrace{\cos(\omega_c t + \omega_m t + 90°)}^{\text{USB}}. \tag{4.4.10}$$

$$v_2 = \cos[\omega_c t - (\omega_m t + 90°)] - \cos[\omega_c t + (\omega_m t + 90°)]$$

$$= \underbrace{\cos(\omega_c t - \omega_m t - 90°)}_{\text{LSB}} - \overbrace{\cos(\omega_c t + \omega_m t + 90°)}^{\text{USB}}. \tag{4.4.11}$$

The output of the adder is given as

$$v_0 = v_1 + v_2 = 2\cos(\omega_c t + \omega_m t + 90°). \tag{4.4.12}$$

It can be shown that if both signals are fed phase-shifted to the balanced modulator, an SSB output with the LSB will be obtained. This is left as an exercise to the reader.

The third method

This method was developed by Weaver as a means of retaining the merits of the phase-shift method without the associated demerit of maintaining an AF phase-shift network over a large range of audio frequencies. This method is very complex and very rarely used commercially. The block diagram of the SSB generator using the *Weaver's method* is given in Figure 4.14, which is self-explanatory.

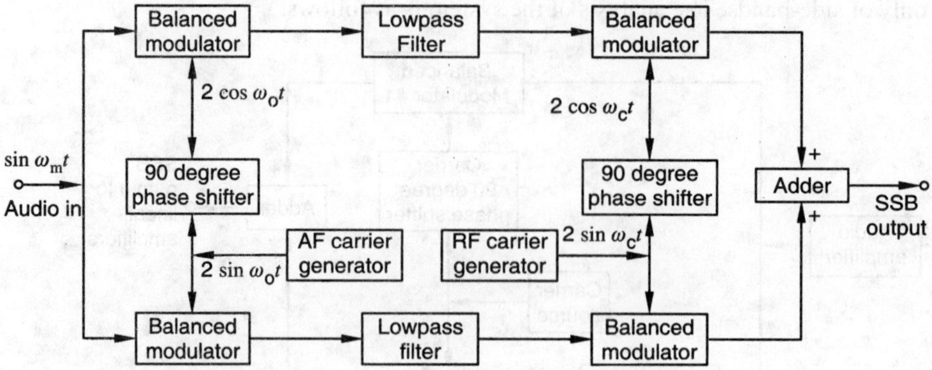

Figure 4.14 *Block diagram of an SSB transmitter using Weaver's method*

4.4.3 SSB reception

Theoretically, it can be shown that it is possible to demodulate SSB signals by multiplying them with a synchronous carrier. Hence, balanced-modulator circuits or product demodulator circuits are used for demodulation. The carrier signal for the demodulator must be locally generated if the signals are true SSB signals with the carrier completely suppressed. This needs extreme stability for the local oscillator in the receiver front-end. Crystal-controlled oscillators are used.

Product demodulator

The product demodulator is virtually a mixer with audio output. It is popular in SSB system, but it is equally capable of demodulating all other forms of AM as well. With reference to Figure 4.15, the input SSB signal is fed to the base through a fixed-frequency IFT (IF transformer), and the signal from a crystal oscillator is applied to the un-bypassed (i.e., in common collector configuration) emitter.

Balanced modulator for SSB detection

This is mainly used in portable SSB receivers, where it is of paramount importance to have as few components as possible and to save weight and power. As the diode balanced modulator can demodulate SSB signals, it is used in transceivers. The circuit of the diode demodulator is given in Figure 4.16.

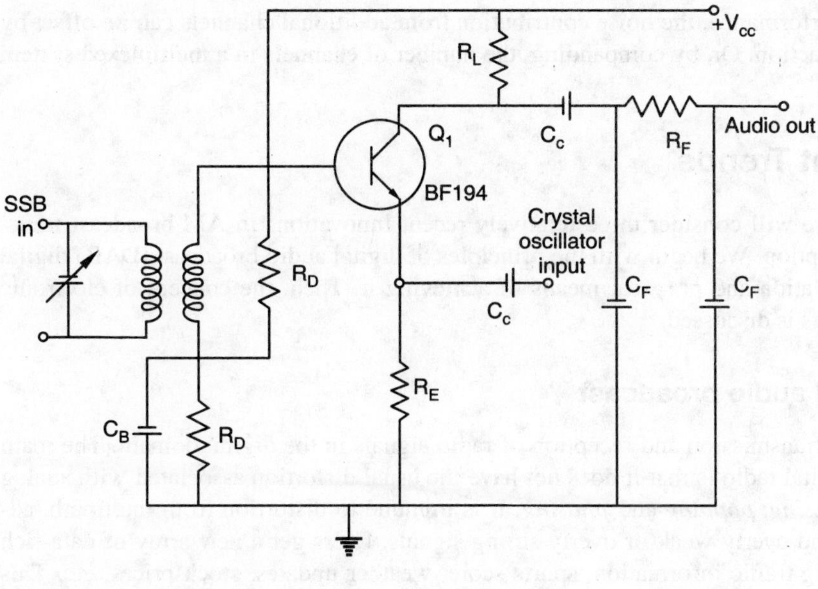

Figure 4.15 *Product demodulator*

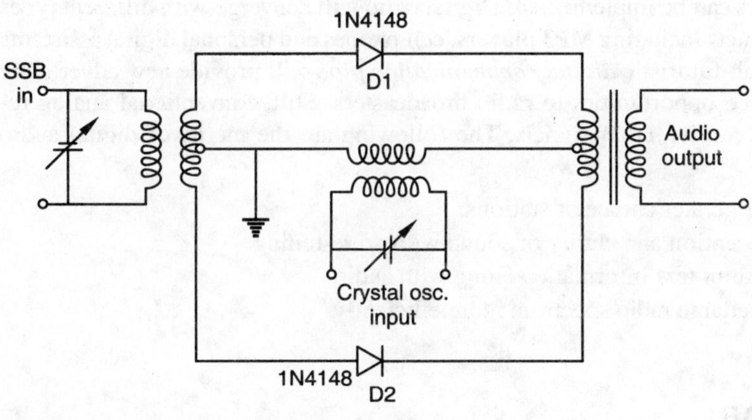

Figure 4.16 *Diode demodulator for SSB*

4.4.4 Companded SSB

Companding is a technique used extensively to combat the effects of channel noise in speech transmission. The speech signal is compressed in volume range relative to a fixed level (typically at – 10dBm). A compression ratio of typically 1 to 2 is applied to the decibel difference between the reference level and the average signal level. At the output of the channel, the signal is expanded by the same ratio to null the effect of compression at the transmitter. The term *compander* is used to refer the unit used to compress and expand, is coined from *compressor–expander*. The main advantage of companding is that it reduces *idle noise* on the channel. *Idle noise* is the background noise heard during the quiet intervals between speech bursts. For a given level of performance, the noise contribution from additional channels can be offset by the companding action. Or, by companding, the number of channels in a multiplexed system can be increased.

4.5 Recent Trends

In this section we will consider three relatively recent innovations in AM broadcast transmission and reception. We begin with the principles of digital audio broadcast (DAB) digital amplitude modulation and practical means of achieving it. Then, the concept of electronic tuning in receivers is discussed.

4.5.1 Digital audio broadcast

Digital radio is transmission and reception of radio signals in the digital domain. The main advantage of digital radio is that it does not have the usual distortion associated with analog radio such as *hissing, popping* and *phasing*. It is immune to distortion from multipath, adjacent stations and overly weak or overly strong signals. Users get a new array of data-rich services including traffic information, sports score, weather updates, stock prices, etc. This data is displayed on a liquid crystal display (LCD) in the form of text, images, and video. Audio features such as time-shift recording or digital recording of playlists using transmitted electronic program gates can be implemented. Digital radio will converge with different types of compelling end products including MP3 players, cell phones and personal digital assistants (PDA). This economical, futuristic *digital communication pipe* will provide new advertising and electronic commerce opportunities to radio broadcasters. Still, conventional analog receivers will be able to receive DAB signals. The following are the merits of digital audio broadcasting:

- DAB gives us far greater choice of stations.
- It offers better reception and clarity of sound with no re-tuning.
- DAB provides useful text information along with audio.
- DAB uses the available radio spectrum more efficiently.
- It is *free-on-air*.

The working of DAB

The various steps involved in DAB can be summarised as:

1. The analog information signal from the local stations and the digital information from the digitised stations are bundled together and transmitted using TDM techniques.
2. Digital signals are compressed to 96 kbps using audio codecs before bundling and transmission.
3. The most common form of interference resulting from multipath reception is removed using *high-definition radio (HD radio)*.

4.5.2 High-definition radio

HD Radio is a new technology that enables AM and FM radio stations to broadcast their programs digitally. The digital broadcasts provide listeners radically improved audio quality and reception and new data services. HD radio enables the reception of CD-quality audio with bandwidth efficiency. HD Radio is a registered trademark of iBiquity Digital Inc.

The benefits of HD Radio are the following:

- FM broadcasts have CD-quality sound.
- AM broadcasts will sound as good as today's analog FM stereo.
- Reception is static-free, without pops, hiss, and fades.
- New data services, such as scrolling text displayed on a radio screen with song titles, artists' names, traffic updates, weather forecasts, sports scores and much more are available.
- HD Radio provide opportunities for more advanced audio services such as *surround sound*, multiple audio sources at the same dial position, on-demand audio services, store-and-replay, overlaying real-time traffic information on a navigational map, and a host of other services.
- No subscription fees
- Easy transition for broadcasters and consumers by using the existing infrastructure and spectrum[4]
- Enhanced sound fidelity
- Improved reception
- Upgraded audio quality

The key players in digital radio markets are the following:

1. Roberts
2. JVC
3. Kenwood
4. Panasonic
5. Intempo
6. Arcam
7. pure-Digital
8. Grundig
9. Cambridge Audio
10. Goodmans

4.5.3 Digital amplitude modulation

Linear amplitude modulation in very high-power, full-carrier systems still requires rather large solid state or vacuum tube devices in medium wave (MW) and shortwave (SW), telephone, VHF and UHF television transmission broadcast systems. The analog format for these systems for MW and SW is a simple full carrier 100% AM with no phase modulation. Full-carrier amplitude modulation is the oldest form of baseband transmission over a carrier signal by radio. The baseband modulation signal usually takes a CW voltage carrier from an unmodulated static level to near cutoff when the modulating signal is at its highest negative voltage value with respect to the carrier polarity, and to twice its static value when it is at its highest positive value with respect to the carrier polarity. The 100% modulation of a carrier in this manner results in an instantaneous carrier voltage that swings up to two times the unmodulated level, down to zero and back again if the peak-to-peak voltage of the modulating signal is equal in magnitude to the carrier voltage. It is not possible to properly modulate the carrier to any

[4] *www.db.com*

levels beyond twice the carrier voltage level or down past zero without phase inversion of the carrier and saturation of the active devices.

There are many ways to realise full-carrier amplitude modulation. Perhaps the earliest used is the high-level modulation of the power output stage operating in saturated class-C. This is done simply by inserting the modulation voltage in series with the amplifier output circuit dc supply voltage. The modulator power capability must be equal to half the carrier power in order to cause the carrier to instantaneously reach twice its unmodulated value and zero. As the instantaneous carrier voltage reaches twice the unmodulated value, the peak envelope power of a 100% modulated full-carrier system is four times the carrier power. The advantage of high-level modulation is that the power RF stage is operated in saturated class C, which can have a power efficiency in the order of 80%. It is also possible to use low-level modulation, where a low power stage in a transmitter is modulated as if it were the high-level stage but then amplified with linear amplifiers to the desired power level. A high degree of linearity is required after modulation so that the modulation envelope is not distorted on its way to becoming high power.

There are serious disadvantages with traditional high- or low-level AM methods not the least of which is linear amplitude distortion in the modulator itself. Active devices used to amplify amplitude modulated signals have nonlinear transfer functions which may result in unacceptable differential gain and phase performance, compression at the top, crossover distortion near cutoff, incidental carrier phase modulation, and cross-modulation, all resulting in the in-band or odd-order intermodulation products. Intermodulation products which are in-band are added signal components which were not present in the original set of modulating signals. Various methods and circuits have been devised to minimise the intermodulation products by pre-correcting or pre-distorting the low-level modulated stage in a way that they counter the distortion caused by follow-on linear amplifiers with a theoretical net effect of zero distortion, but success is usually limited to about 10 dB of correction. Clearly, in this age of wireless communication demanding a much higher degree of linearity, other methods of providing high-level linear amplification must be sought, and are possible.

The 3 dB doupler

The 3 dB hybrid used as a combiner is the heart of the modulator in the digital amplitude modulation. Referring to Figure 4.17, a wave of power P_{in1}, travelling from left to right enters the combiner at a relative phase of zero degrees (arbitrarily chosen to be the phase reference) and undergoes a change to $-90°$ through the quarter wavelength path distance to P_{out}. A second wave of equal power already set to $-90°$ entering at P_{in2} will add completely with P_{in1} at P_{out} as it is coherent with P_{in1} at the same cross-section point along the line. Power inputs to this four-port device do not have to be equal in magnitude, nor do they have to be 90° out of phase with each other, but in the derivation of a digital amplitude modulator, phase will always be assumed to have proper directional shift of a constant 90° so as to cause the power addition to be a function only of the relative amplitudes of the two signals being summed. In the general case,

$$P_{\text{out}} = \left(\sqrt{\frac{P_{\text{in1}}}{2}} + \sqrt{\frac{P_{\text{in2}}}{2}}\right)^2 \tag{4.5.13}$$

and,

$$P_R = \left(\sqrt{\frac{P_{\text{in1}}}{2}} - \sqrt{\frac{P_{\text{in2}}}{2}} \right)^2 \tag{4.5.14}$$

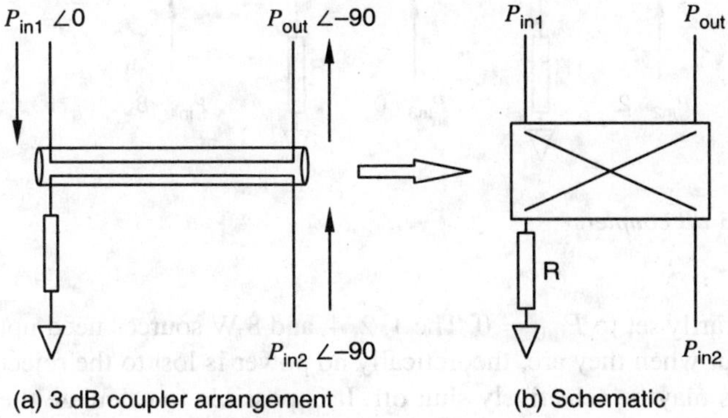

Figure 4.17 *A 3 dB coupler*

For example, let $P_{\text{in1}} = P_{\text{in2}} = 2$. It is easy to show that above equations give a P_{out} of 4 and the power dissipated in the resistor, P_{R} , sometimes called the reject load, of 0.

A cascade of combiners

In Figure 4.18, a cascade of four combiners is shown.

For a real example, arbitrarily let $P_{\text{in0}} = P_{\text{in1}} = 1$ W, $P_{\text{in2}} = 2$ W, $P_{\text{in3}} = 4$ W and $P_{\text{in4}} = 8$ W respectively. No power is wasted in the reject loads because both inputs to each combiner are always equal in power level with proper 90° phase shift. They carry power coming from the imaginary infinite array of combiners to the left represented by 1 watt in the example shown in Figure 4.18. There is no control over it as the carrier power as a dynamic source does not exist. It is always either present or absent, 1 W or 0 or any other constant

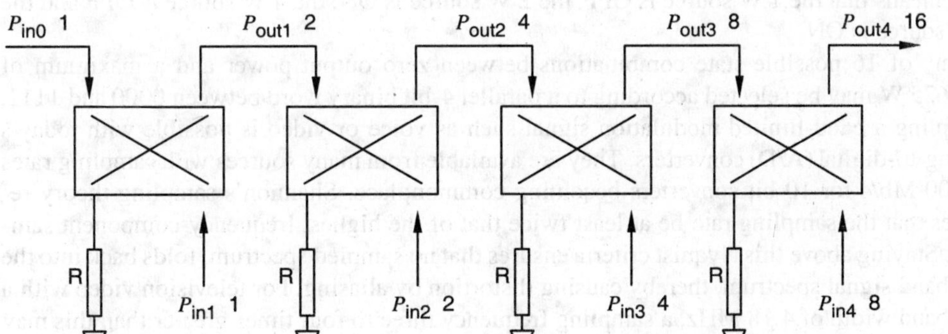

Figure 4.18 *A cascade of 3 dB couplers*

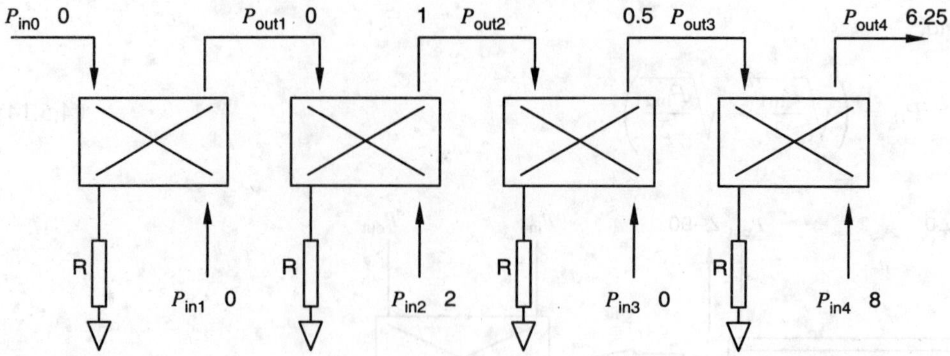

Figure 4.19 *A cascade with four 3 dB couplers*

value that makes sense. It is arbitrarily set to $P_{in0} = 0$. The 1, 2, 4, and 8 W sources need not all be present at the same time, but when they are, theoretically no power is lost to the reject loads. Some of the power sources may be selectively shut off. It is possible to calculate the power that will be at the output port of the cascade. In Figure 4.19, the first and the third power sources are set to 0. It is expected that some power will be lost in the reject loads as power levels into an individual combiner are no longer equal. The power at the output of the cascade has dropped dramatically from the 14.0625 relative watts which would have been obtained if all the 'carrier sources' are present as in Figure 4.19. In fact, the maximum that the output power can ever be is 14.0625 when all the sources are ON, and the lowest is zero when all the sources are OFF.

Digital amplitude modulator

From the least significant power position (least significant bit or LSB) to the left in Figure 4.19 to the highest significant power position (most significant bit or MSB) to the right, a parallel binary word may be derived to be 0101. (To be true to the proper order of writing numbers, the MSB should be to the left so that 0101 becomes 1010, but this is only a convention and the order is chosen to remain 0101 as it conforms to the flow of power addition in the Figure). This means that the 1 W source is OFF, the 2 W source is ON, the 4 W source is OFF and the 8 W source is ON.

Any of 16 possible state combinations between zero output power and a maximum of 14.0625 W may be selected according to a parallel 4-bit binary word between 0000 and 1111. Sampling a band-limited modulation signal such as voice or video is possible with today's analog-to-digital (A/D) converters. They are available from many sources with sampling rates of 100 Mb/s for 10-bit converters becoming commonplace. Shannon's sampling theory requires that the sampling rate be at least twice that of the highest frequency-component sampled. Staying above this Nyquist criteria ensures that no sampled spectrum, folds back into the baseband signal spectrum, thereby causing distortion by aliasing. For television video with a baseband width of 4.18 MHz, a sampling frequency three to four times greater than this may easily be chosen and implemented. For the moment it is sufficient to understand that the baseband signal must be band-limited so that the minimum sampling frequency is more than twice the bandwidth.

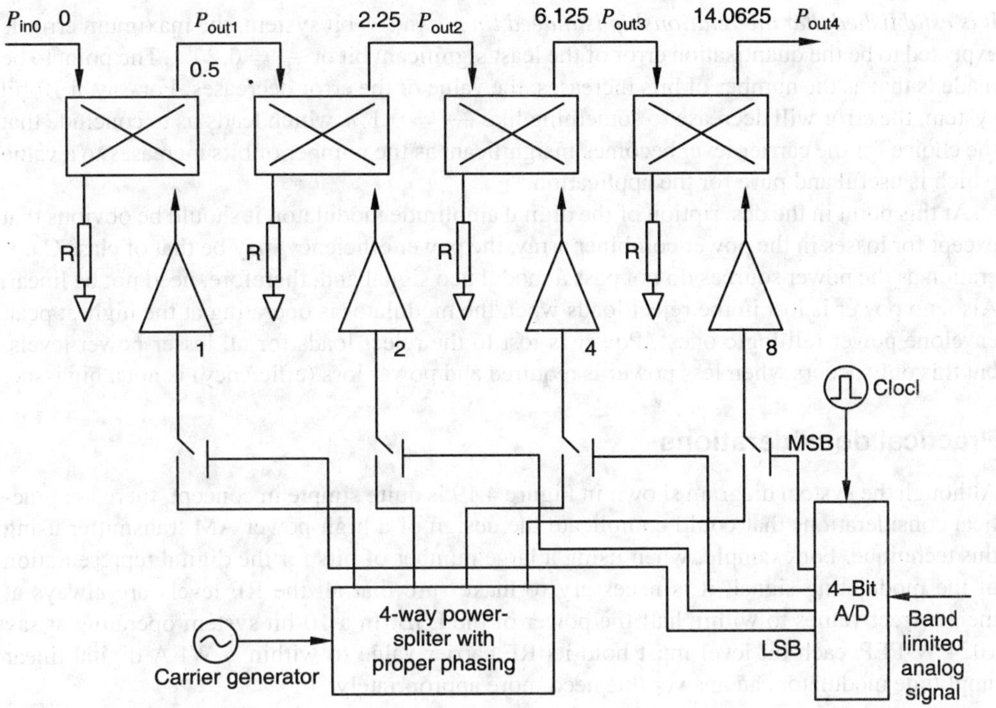

Figure 4.20 *Block diagram of digital amplitude modulator*

The schematic block diagram of a digital amplitude modulator is shown in Figure 4.20. The highest value for the analog signal waveform is assumed to be 16 volts and the lowest is 0. The signal is fed into an A/D converter capable of sampling such a waveform such that, when the analog signal reaches its peak of 16 V, all output bits at the parallel output are logic ones, i.e. the output is 1111. When the modulating signal is at its lowest value, 0, all output bits are logic zeros, i.e. the output is 0000. This is a voltage waveform with all other in between values properly represented by the correct binary word as determined by the A/D converter. For example, if, at the moment of sampling the analog waveform is at an 8 V point, the digital word at the output of the A/D converter is 0001 (in keeping with the chosen convention of MSB at the rightmost place). If it is at a 5 V point, then the bits change to 1010. This straight binary representation of the analog sampled point may be used to control an RF drive switch to each corresponding power source feeding the combiner array. If a bit is logic 1, the drive switch is closed. If it is logic 0, it is open.

It has been shown that when the modulating signal is at its maximum value, i.e., 16 V in this example, all bits in the 4-bit word are logic 1 (1111), which makes the power output of the combiner array equal to 14.0625 W. It has also been shown that when the modulating signal is at its minimum, all the bits in the 4-bit binary word are logic 0 (0000), which makes the power output of the combiner array equal to 0.00 W. It should be obvious that when the word is a mix of ones and zeros, the power output will be something in between these extremes. The remaining issue is whether or not there is a linear relationship between the sampled baseband binary value and the RF voltage output derived as the square-root of the output power. *And, it is established that the relationship is indeed linear.* In a 4-bit system, the maximum error is expected to be the quantisation error of the least significant bit or $\frac{1}{16} = 6.25\%$. The point to be made is that as the number of bits increases, the value of the error decreases. For, say, a 10-bit system, the error will decrease to something like $\frac{1}{2^{10}} = 0.1\%$, which leads us to conclude that the choice for the carrier level becomes insignificant as the number of bits increases to a value which is useful and pure for the application.

At this point in the description of the digital amplitude modulator, it should be obvious that except for losses in the power combiner array, the power efficiency may be that of class C operation as the power sources do not pass a modulated signal and, therefore, need not be linear. Also, no power is lost in the reject loads when the modulator is operating at the highest peak envelope power (all logic ones). Power is lost to the reject loads for all lesser power levels, but this only occurs when less power is required and power loss (efficiency) is not a big issue.

Practical donsiderations

Although the system diagram shown in Figure 4.19 is quite simple in concept, there are practical considerations that could complicate the design of a high-power AM transmitter using this technique. For example, when using a large number of bits for the digital representation of the modulating signal, it is necessary to make sure that all the RF levels are always at their correct values to within half the power of the LSB. In a 10-bit system operating at say 1024 W PEP, each bit level must hold its RF carrier value to within $\frac{1}{2}$ W! A digital linear amplitude modulator can answer this need more appropriately.

4.5.4 Digital Radio Mondale (DRM)

The DRM on-air system will propel the AM broadcasting bands below 30 MHz (short wave, medium wave and long wave) to the next level. It is the only universal, non-proprietary digital AM radio system with near-FM quality sound available to markets worldwide. The quality of DRM audio is excellent, and the improvement upon analog AM is immediately noticeable. DRM can be used for a range of audio content, including multilingual speech and music. Besides providing near-FM quality audio, the DRM system has the capacity for integrating data and text. This additional content can be displayed on DRM receivers to enhance the listening experience.

Unlike digital systems that require a new frequency allocation, DRM uses existing AM broadcast frequency bands. The DRM signal is designed to fit in with the existing AM broadcast band plan, based on signals of 9 kHz or 10 kHz bandwidth. It has modes requiring as little as 4.5 kHz or 5 kHz bandwidth, plus modes that can take advantage of wider bandwidths such as 18 or 20 kHz.

Many existing AM transmitters can be easily modified to carry DRM signals. DRM applications will include fixed and portable radios, car receivers, software receivers, and PDAs. Several early prototype DRM receivers have been produced, including a software receiver. The DRM system uses a type of transmission called COFDM (Coded Orthogonal Frequency Division Multiplex). This means that all the data, produced from the digitally encoded audio and their associated data signals, is shared out for transmission across a large number of closely spaced carriers. All of these carriers are contained within the allotted transmission channel. The DRM system is designed so that the number of carriers can be varied, depending on factors such as the allotted channel bandwidth and degree of robustness required.

The DRM system can use three different types of audio coding, depending on the broadcaster's preferences. MPEG4 AAC audio coding, augmented by SBR bandwidth extension, is used as a general-purpose audio coder and provides the highest quality. MPEG4 CELP speech coding is used for high quality speech coding where there is no musical content. HVXC speech coding can be used to provide a very low bit-rate speech coder. The robustness of the DRM signal can be chosen to match different propagation conditions.

4.5.5 Electronically tuned receivers

Most of the modern receivers have electronic digital tuning systems which are based on computer control of varactor tuning diodes. This is especially true for automotive receivers, where a microcomputer controller takes over many of the scanning and tuning functions, thus freeing the driver from distractions. Several integrated circuit manufacturers are producing ICs that can take care of the electronic tuning function. National Semiconductor's LM1863 is one such chip, which contains the local oscillator, mixer, IF amplifiers and detectors in a single chip. Another chip is Motorola's MC13025P, an electronically tuned radio front-end. The block diagram of the above chip is given in Figure 4.21. The MC13025P provides a high dynamic range mixer, voltage controlled oscillator and the first IF, which, with the MC13022 and synthesiser forms a complete digitally controlled AM stereo tuner system. This system in turn may drive a dual-channel audio processor and high-power amplifiers for car radio or home stereo applications. Other applications include portable radio *boom boxes*, table radios, and component stereo systems.

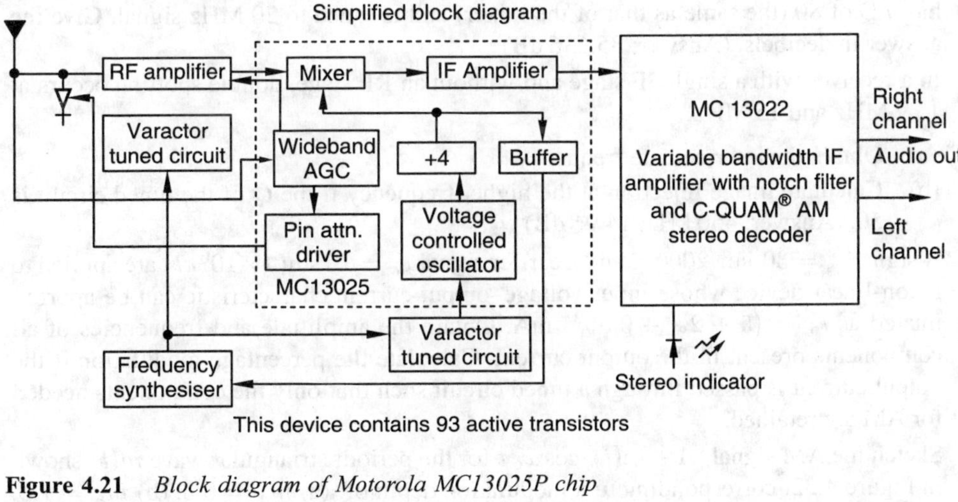

Figure 4.21 *Block diagram of Motorola MC13025P chip*

Exercises

1. A receiver is tuned to the 3 to 30 MHz high-frequency band in one range using an IF of 40.525 MHz. Find the range of oscillator frequencies, the range of image frequencies

and the type of filter needed to make the receiver function properly. (Answer: 43.525 MHz to 70.525 MHz, 84.05 MHz to 114.05 MHz, bandpass filter with centre frequency 40.525 MHz.)

2. A superheterodyne receiver tunes to the range from 4 to 10 MHz with an IF of 1.8 MHz.
   (a) Find the image frequency range.
   (b) If the RF circuits have combined effective Q of 50 at the top of this band, find the image rejection ratio in dBs at that frequency. (Answer: 7.6 MHz to 13.6 MHz, 29.897 dB)
3. When a superheterodyne receiver is tuned to 760 kHz, its local oscillator provides the mixer with an input of 1215 kHz.
   (a) What is its image frequency?
   (b) The antenna of this receiver is connected to the mixer via a tuned circuit whose loaded Q is 50. What will be the rejection ratio for the calculated image frequency? (Answer: 1670 kHz, 38.80 dB)
4. Calculate the image rejection of a receiver having an RF amplifier and an IF of 455 kHz if the Qs of the relevant coils are 70 at an incoming frequency of (a) 1300 kHz, (b) 30 MHz. (Answer: 75.646 dB, 25.347 dB)
5. A superheterodyne receiver having an RF and an IF of 455 kHz is tuned to 20 MHz. Calculate the Qs of the RF and mixer stages if the receiver's image rejection is to be 140. (Answer: 132.44 each)
6. Calculate the image frequency rejection of a double-conversion receiver which has a first IF of 4 MHz and a second IF of 400 kHz, an RF amplifier whose tuned circuit has a Q of 80 (the same as that of the mixer) and is tuned to 20 MHz signal. Give the answer in decibels. (Answer: 85.636 dB)
7. In a receiver with a single IF stage and without an RF stage, double spotting occurs at 15.9 MHz and 15 MHz.
   (a) Obtain the intermediate frequency
   (b) Calculate image rejection at the higher frequency if the Q of the tuned circuit is 50. (Answer: 450 kHz, 14.96 dB)
8. A signal $v_s = 30 \sin 2000\pi t$ and a carrier signal $v_c = 10 \sin(4 \times 10^6 \pi t)$ are applied to a non-linear device whose input-voltage–output-current characteristic can be approximated as $i_o = (5 + 2v_i - 0.5v_i^2)$ mA. Obtain the amplitude and frequencies of all components present in the output current. Calculate the percentage modulation if the output current is passed through a tuned circuit such that only the components needed for AM are retained.
9. Sketch the AM signal $[A + m(t)] \cos \omega_c t$ for the periodic triangular wave $m(t)$ shown in Figure 4.22. corresponding to a modulation depth of: (a) $m_a = 0.5$, (b) $m_a = 1.0$. Also, plot using MATLAB the AM wave for these two cases.
10. If the input to the SSB system in Figure 4.13 is $m_h(t)$ instead of $m(t)$, what will be the output? Is this signal still an SSB signal? Can it be demodulated back to $m(t)$?
11. Refer to the cascade of four 3 dB couplers shown in Figure 4.20 as part of the digital amplitude modulator. Assume that binary code given at switches from the A/D converter is 1010 (MSB at the right most position).

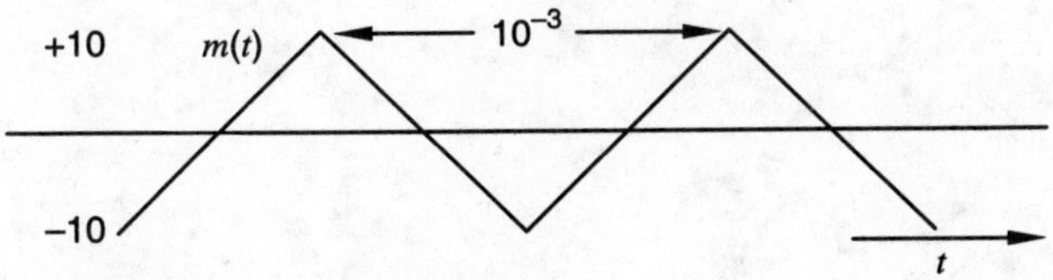

Figure 4.22 *A triangular AM*

(a) Calculate the net output power if the power at the left-most combiner is 2 W.

(b) What is the percentage loss in power compared to full power on input situation?

(c) What percentage of input power is dissipated in the resistors? (Answer: 3.125 W, $\frac{25}{28.125} = 88.89\%$, $\frac{6.875}{10} = 68.75\%$)

12. Repeat the above problem for a 10-bit A/D system when the binary input is 01010 10101 (MSB at rightmost position) and the reference power is 2 W.(Answer: 908.445 W, $\frac{1135.56}{2044.0} = 55.56\%$, $\frac{1364-908.445}{1364} = 33.40\%$)

Chapter 5

Angle Modulation

In the previous chapters, we have been discussing amplitude modulation (AM) and its various flavours, all of which can be classified as *linear modulation* schemes. The nomenclature is so chosen to indicate that the modulated signal can be obtained by applying the *principle of superposition* on the modulating signals. In this chapter we look into another form of modulation wherein the carrier frequency is modulated rather than its amplitude. The resulting modulation is called *frequency modulation (FM)*. We can also modulate the phase of the carrier; the resulting modulation is termed as *phase modulation*. These two modulation schemes, viz. FM and PM, are collectively known as *angle modulation* schemes. Note that both of them are nonlinear modulation schemes. Hence, we cannot apply the *principle of superposition* to obtain the modulated signal from the modulating components. We begin our discussion with FM.

5.1 Frequency Modulation

In FM, the frequency of the carrier wave is changed according to the instantaneous value of the modulating signal. In this process the carrier amplitude remains constant. The amplitude of the modulating signal will cause the carrier frequency to deviate from its normal frequency (i.e., its frequency without any modulation) by a certain amount. As per specifications of the FCC, the maximum frequency deviation allowed is 75 kHz. The shift in the carrier frequency from its resting point with respect to the maximum frequency of the modulating voltage is called the *deviation ratio*, for which the maximum value stipulated by the FCC is 5.

$$\text{Deviation ratio} = \frac{f_{\text{dev(max)}}}{f_{\text{AF(max)}}},$$

where the maximum frequency deviation of the carrier is $f_{\text{dev}}(\text{max})$ and the frequency of the corresponding modulating signal is f_{AF}. The *modulation index* is defined as

$$\text{Modulation index, } m_{\text{f}} = \beta = \frac{f_{\text{dev}}(\text{max})}{f_{\text{AF}}}.$$

5.1.1 Frequency spectrum of FM

The frequency spectrum of the FM wave can be derived as discussed below.

Let us assume that the carrier wave is represented by $e_c = E_c \cos \omega_c t$ and the modulating signal by $e_m = E_m \cos \omega_m t$. The instantaneous value of carrier frequency on modulation, f, can be represented as

$$f = f_c + kE_m \cos \omega_m t, \tag{5.1.1}$$

where

$$\begin{aligned} f_c &= \text{unmodulated carrier frequency} \\ k &= \text{proportionality constant} \\ E_m \cos \omega_m t &= \text{instantaneous modulating signal.} \end{aligned}$$

The maximum deviation for this particular carrier will occur when the cosine term has its maximum value ± 1. Now, the instantaneous frequency will be

$$f = f_c + kE_m \cos \omega_m t.$$

Hence, the maximum deviation in frequency δ is given by

$$\delta = \pm kE_m. \tag{5.1.2}$$

The instantaneous amplitude of the FM signal is given by

$$e = E_c \cos \theta. \tag{5.1.3}$$

Now,

$$\begin{aligned} \omega &= 2\pi f = \omega_c + 2\pi kE_m \cos \omega_m t \\ \theta &= \int \omega dt = \int (\omega_c + 2\pi\, kE_m \cos \omega_m t) dt = \left(\omega_c t + \frac{2\pi\, kE_m \sin \omega_m t}{\omega_m}\right) \\ &= \omega_c t + \frac{kE_m \sin \omega_m t}{f_m} = \omega_c t + \frac{\delta}{f_m} \sin \omega_m t. \end{aligned} \tag{5.1.4}$$

Substituting the value of θ in (5.1.3), we get the instantaneous value of the FM signal. Hence,

$$e = E_c \cos \left(\omega_c t + \frac{\delta}{f_m} \sin \omega_m t\right). \tag{5.1.5}$$

The *modulation index* for FM, β is defined as

$$\beta = \frac{\text{maximum frequency deviation}}{\text{modulating frequency}} = \frac{\delta}{f_m}. \tag{5.1.6}$$

Substituting in (5.1.5), we get

$$e = E_c \cos (\omega_c t + \beta \sin \omega_m t). \tag{5.1.7}$$

It can be shown that for a sinusoidal modulating signal, assuming that e represents the instantaneous voltage impressed across a 1Ω resistor, the average power over a cycle of the modulating frequency is given by

$$\frac{1}{T}\int_0^T e^2\,dt = \frac{E_c^2}{T}\int_0^T \cos^2(\omega_c t + \beta\sin\omega_m t)\,dt \tag{5.1.8}$$

where $T = \frac{1}{f_m}$. This expression may be rewritten as

$$\frac{E_c^2}{T}\int_0^T \frac{1+\cos(2\omega_c t + 2\beta\sin\omega_m t)}{2}\,dt.$$

The second part of the integral is zero since it is periodic in T. This assumes $\omega_m = 2\pi/T \ll \omega_c$. The first term gives $\frac{E_c^2}{2}$. Although shown only for a sinusoidal modulating signal, the above result is true for any modulating signal whose highest frequency component B Hz is small compared to the carrier frequency f_c. The reader is advised to prove this.

We will now try to determine the frequency components and their amplitudes for the frequency modulated signal of arbitrary β and sinusoidal modulating signal. Note that we can write equation (5.1.7) as

$$e = E_c\left[\cos\omega_c t\cos(\beta\sin\omega_m t) - \sin\omega_c t\sin(\beta\sin\omega_m t)\right]. \tag{5.1.9}$$

We can expand both $\cos(\beta\sin\omega_m t)$ and $\sin(\beta\sin\omega_m t)$ in their respective Fourier series. Both series may be found simultaneously by considering the periodic complex exponential:

$$v(t) = \exp(j\beta\sin\omega_m t), \quad -\frac{T}{2} < t < \frac{T}{2}. \tag{5.1.10}$$

The real part of this function gives us our cosine function and the imaginary part the sine function. If we expand this exponential in its Fourier series, we can expect to get a real part consisting of even harmonics of ω_m and an imaginary part consisting of the odd harmonics. By equating reals and imaginaries, we shall then obtain the desired Fourier expansions for cos $(\beta\,\omega_m t)$ and sin $(\beta\,\omega_m t)$, respectively.

The Fourier coefficient of the complex Fourier series for the exponential of equation (5.1.10) is given by

$$C_n = \int_{-T/2}^{T/2} \exp\left[j(\beta\sin\omega_m t - \omega_n t)\right]dt; \quad \omega_m = \frac{2\pi}{T}, \quad \omega_n = \frac{2\pi\,n}{T} = n\,\omega_m. \tag{5.1.11}$$

Normalising this integral by letting $x = \omega_m t$, we get

$$\frac{C_n}{T} = \frac{1}{2\pi}\int_{-\pi}^{\pi} \exp\left[j(\beta\sin x - n\,x)\right]dx. \tag{5.1.12}$$

This integral can be evaluated only as an infinite series and is called the *Bessel function of the first kind* and is denoted by the symbol $J_n(\beta)$.

Specifically,

$$J_n(\beta) = \frac{1}{2\pi}\int_{-\pi}^{\pi} \exp\left[j(\beta \sin x - n\, x)\right]\, dx \tag{5.1.13}$$

so that $C_n = T\, J_n(\beta)$.

For $n = 0$, we get $C_0 = T\, J_0(\beta)$, the dc component of the Fourier series representation of the periodic complex exponential of equation (5.1.10). Increasing values of n give the corresponding Fourier coefficients for the higher frequency terms of the Fourier series. The spectrum of the complex exponential of equation (5.1.10) (and ultimately that of the frequency modulated signal) will thus be given by the value of the Bessel function and will depend on the parameter β. Hence

$$\exp\left(j\beta\, \omega_m t\right) = \frac{1}{T}\sum_{n=-\infty}^{\infty} C_n \exp\left(j\omega_n t\right) = \sum_{n=-\infty}^{\infty}.$$

$$J_n(\beta) \exp\left(j\omega_n t\right) \quad \omega_n = n\omega_m. \tag{5.1.14}$$

From equation (5.1.14), the Fourier series expansion of the complex exponential, we can obtain the desired Fourier series for cos (β sin $\omega_m t$) and sin (β sin $\omega_m t$). It can be shown, either from the integral definition of $J_n(\beta)$ or from the power series expansion of $J_n(\beta)$, that

$$J_n(\beta) \quad = \quad J_{-n}(\beta) \quad n \text{ even} \tag{5.1.15}$$

$$\text{and } j_n(\beta) \quad = \quad -J_{-n}(\beta) \quad n \text{ odd}$$

Writing out the Fourier series term by term and using equation (5.1.15) to combine the positive and negative terms of equal magnitude of n, we get

$$\begin{aligned}\exp(j\beta \sin \omega_m t) \quad &= \quad J_0(\beta) + 2\left[J_2(\beta) \cos\, 2\omega_m t + J_4(\beta) \cos\, 4\omega_m t + \ldots\right] \\ &\quad +2j\left[J_1(\beta) \sin\, \omega_m t + J_3(\beta) \sin\, 3\omega_m t + \ldots\right].\end{aligned} \tag{5.1.16}$$

But

$$\exp\left(j\beta\, \omega_m t\right) \quad = \quad \cos\left(\beta \sin\, \omega_m t\right) + j \sin\left(\beta\, \omega_m t\right). \tag{5.1.17}$$

Equating the real and imaginary terms, we get:

$$\begin{aligned}\cos(\beta \sin\, \omega_m t) \quad &= \quad J_0(\beta) + 2J_2(\beta) \cos\, 2\omega_m t + \\ &\quad 2J_4(\beta) \cos\, 4\omega_m t + \ldots\end{aligned} \tag{5.1.18}$$

$$\text{and } \sin(\beta \sin\, \omega_m t) \quad = \quad 2J_1(\beta) \sin\, \omega_m t + 2J_3(\beta) \sin\, 3\omega_m t + \ldots \tag{5.1.19}$$

Now using the Fourier series expansions for the cosine and sine terms, and then utilising the trigonometric sum and difference formulae, we get

$$
\begin{aligned}
e &= E_c\{J_0(\beta)\cos\omega_c t \\
&+ J_1(\beta)[\cos(\omega_c+\omega_m)t - \cos(\omega_c-\omega_m)t] \\
&+ J_2(\beta)[\cos(\omega_c+2\omega_m)t + \cos(\omega_c-2\omega_m)t] \\
&+ J_3(\beta)[\cos(\omega_c+3\omega_m)t - \cos(\omega_c-3\omega_m)t] \\
&+ J_4(\beta)[\cos(\omega_c+4\omega_m)t + \cos(\omega_c-4\omega_m)t]\ldots\}.
\end{aligned}
\tag{5.1.20}
$$

It can be seen that the FM output signal consists of a carrier and an apparently infinite number of sidebands, each preceded by J coefficients. $J_n(\beta)$ may be shown to be a solution of an equation of the form

$$(\beta)^2\frac{d^2y}{d\beta^2} + \beta\frac{dy}{d\beta} + (\beta^2 - n^2)y = 0. \tag{5.1.21}$$

The solution, i.e., the formula for the Bessel function is

$$J_n(\beta) = \left(\frac{\beta}{2}\right)^n\left[\frac{1}{n!} - \frac{(\beta/2)^2}{1!(n+1)!} + \frac{(\beta/2)^4}{2!(n+2)!} - \frac{(\beta/2)^6}{3!(n+3)!} + \ldots\right]. \tag{5.1.22}$$

Yet another form of the Bessel function is[1]

$$J_n(\beta) = \beta^n\sum_{m=0}^{\infty}\frac{(-1)^m\beta^{2m}}{2^{2m+n}m!(n+m)!}. \tag{5.1.23}$$

The MATLAB function *bessfn.m* that created the plots in Figure 5.1 is listed below.

```
function bessfn(n,mf)
% Computes and plots Bessel Functions of the
% First kind up to order n.
% Modulation Index varies from 0 to mf.
m=[0:.01:mf];
for i=1:n
    bf=besselj(i-1,m);
    plot(m,bf); hold on; grid on;
end;
hold off;
```

We can make the following observations on the FM signal.

1. Unlike the case of AM, *FM signal contains an infinite number of sidebands* as well as the carrier. They are separated from the carrier by $\pm f_m$, $\pm 2f_m$, $\pm 3f_m$, ..., and thus have a recurrence frequency of f_m.
2. The J coefficients eventually decrease in value as n increases. As is seen from Figure 5.1, the value fluctuates on either side of zero, gradually diminishing.

[1] A detailed study on Bessel functions can be found in references [2] and [14].

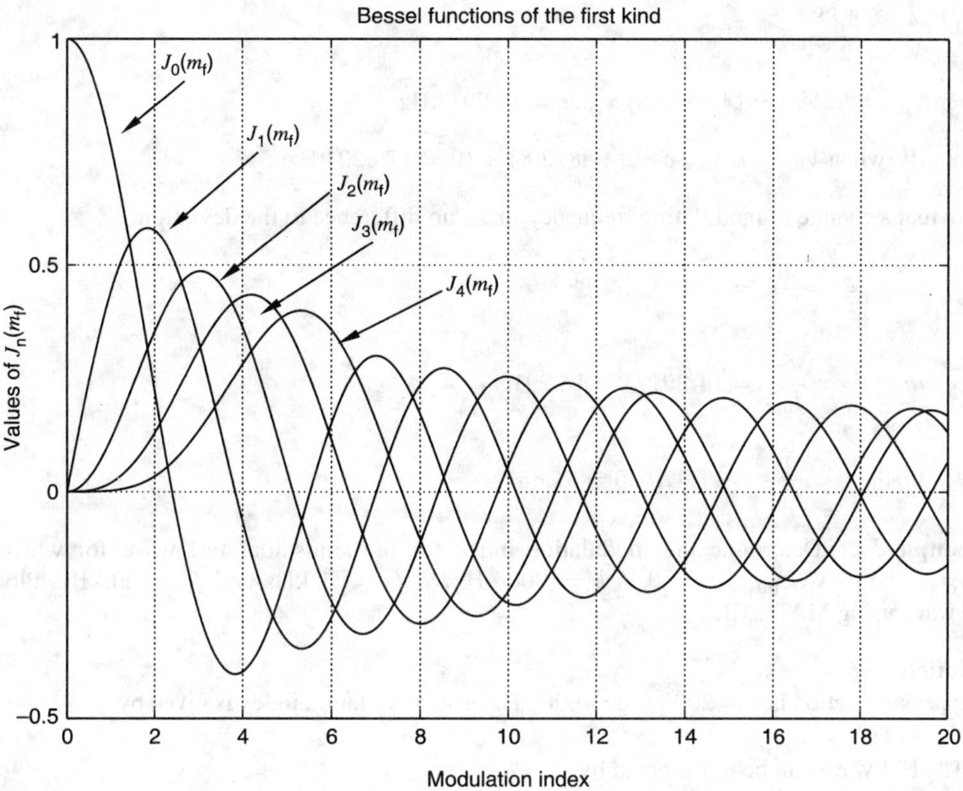

Figure 5.1 *Bessel functions*

3. The modulation index determines how many sideband components have significant amplitude.
4. The sidebands at equal distance from f_c have equal amplitudes. The J coefficients have negative values, occasionally, signifying a 180° phase change for that pair of sidebands.
5. In FM the total transmitted power remains constant, but with increased depth of modulation the required bandwidth is increased.
6. In FM the amplitude of the carrier component does not remain constant. Its J coefficient is J_0, which is dependent on β.
7. It is possible for the carrier component to disappear completely. This happens for certain values of modulation index called *eigen values*. This is evident from the plot for $J_0(\beta)$ in Figure 5.1, where it becomes zero for the values of β equal to 2.4, 5.5, 8.6, 11.8 and so on.

Example 5.1 In an FM system, when the AF is 1000 Hz and the AF voltage is 2.8 V, the deviation is 4.88 kHz. If the AF voltage is now increased to 8.2 V, what is the new deviation? If the AF voltage is raised to 10 V while the AF is dropped to 500 Hz, what is the deviation? Also calculate the modulation index in each case.

Solution

As $\delta \propto V_m$, we have

$$\frac{\delta}{V} = \frac{4.88}{2.8} = 1.743.$$

When $V_{\mathrm{m}} = 8.2V, \delta = (4.88/2.8) \times 8.2 = 14.291$ kHz.

Similarly, when $V_{\mathrm{m}} = 10V, \quad \delta = (4.88/2.8) \times 10 = 17.429$ kHz.

Note that a change in modulating frequency make no difference to the deviation.

$$m_{f1} = \frac{\delta_1}{f_{m1}} = 4.88/1 = 4.88.$$

$$m_{f2} = \frac{\delta_2}{f_{m2}} = 14.291/1 = 14.291.$$

$$m_{f3} = \frac{\delta_3}{f_{m3}} = 17.429/.5 = 34.857.$$

Example 5.2 Determine the modulation index for the sinusoidal FM wave for which $E_{\mathrm{c(max)}} = 10$ V, $E_{\mathrm{m(max)}} = 3$ V, $k = 2000$ Hz/V, $f_{\mathrm{m}} = 1$ kHz and $f_{\mathrm{c}} = 20$ kHz. Plot the wave using MATLAB.

Solution
The peak deviation is $\delta = 2000 \times 3 = 6000$ Hz. The modulation index is given by $\beta = \frac{\delta}{f_{\mathrm{m}}} = 6/1 = 6$

The FM wave can be represented by

$$e_{\mathrm{FM}} = 10 \sin\left[40000\pi\, t + 6 \sin\left(2000\pi\, t\right)\right].$$

The following MATLAB code can plot the FM wave along with the modulating signal, $e_{\mathrm{m}} = 3 \cos\left(2000\pi\, t\right)$.

```
% MATLAB code to plot an FM wave..
% fc= 20kHz and fm=1kHz
% mf=6
t=[0:.000001:.004];
efm=10*sin(40000*pi*t+6*sin(2000*pi*t));
plot(t,3*cos(2000*pi*t)); axis([0 .004 -10 10]);hold on;
plot(t,efm);hold off;
%%
```

The plot of the FM wave is given in Figure 5.2.

5.1.2 FM radio frequency band

The range of frequencies from 88 MHz to 108 MHz was set aside by the FCC for FM radio broadcasting in 1945. Each station in this band is allowed ± 75 kHz deviation (150 kHz carrier swing); a 25 kHz guardband is then added above and below the carrier frequency swing. Thus, a total bandwidth of 200 kHz is assigned to each station in this frequency band. There

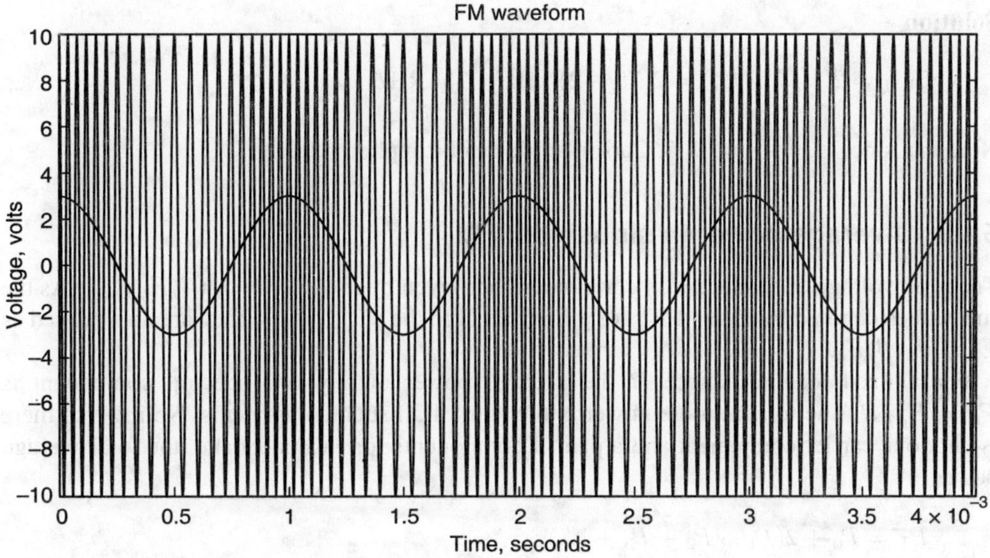

Figure 5.2 *Frequency modulated waveform*

are a total of 100 such FM channels in the FM broadcast band. The FM channel numbering system starts at channel 201 and ends at channel 300. Channel 201 has a carrier rest frequency of 88.1 MHz while the next channel 202 has a carrier rest frequency of 88.3 MHz. The first 20 channels in this band are reserved for educational broadcasts (non-profit), and the remaining 80 channels are for commercial broadcast services. The educational channels have a maximum ERP[2] of 10 W.

5.1.3 Bandwidth for FM

It can be seen that, as modulation depth increases, so does the bandwidth. Reduction in modulation frequency increases the number of sidebands, though not necessarily the bandwidth. Although the number of sideband components is theoretically *infinite*, in practice many of sidebands have insignificant relative amplitudes. The exclusion of these sideband components will not distort the modulated wave unduly.

5.1.4 Carson's rule on FM bandwidth

Carson's rule states that the bandwidth required in an FM system is given by

$$BW_{\mathrm{FM}} = 2 \times (\delta + f_{\max}) = 2 \times (\beta + 1) f_{\max}, \qquad (5.1.24)$$

where, δ is the deviation and $f_{\max}$ is the highest modulating frequency. The Carson's rule gives a fairly accurate result if the modulation index is above 6.

Example 5.3 For an FM system the peak deviation is 75 kHz when the modulating frequency is 2 kHz. Obtain the bandwidth required.

[2]The *effective radiated power* (ERP) is the product of the antenna input power and the antenna gain.

Solution

$$B_{FM} = 2 \times (\delta + f_m) = 2(75 + 2) \text{ kHz} = 154 \text{ kHz}.$$

Note that since $\beta = \dfrac{75}{2} = 37.5$, Carson's rule can be applied safely.

5.1.5 Average power in sinusoidal FM

The peak voltage of the spectrum components are given by $E_{n(max)} = J_n(\beta)E_{c(max)}$. As the rms values denoted by E_n and E_c are proportional to the peak values, they are also related as $E_n = J_n(\beta)E_c$.

For a fixed load resistance R the average power of any one spectral component is $P_n = E_n^2/R$. The total average power is the sum of all such components. Noting that there is only one carrier component and a pair of components for each sideband, the total average power is

$$P_T = P_0 + 2(P_1 + P_2 + P_3 + \ldots).$$

In terms of rms voltages this becomes

$$P_T = \frac{E_0^2}{R} + \frac{2}{R}(E_1^2 + E_2^2 + \ldots). \tag{5.1.25}$$

In terms of the unmodulated carrier and the Bessel function coefficients, this is

$$\begin{aligned} P_T &= \frac{E_c^2 J_0^2(\beta)}{R} + \frac{2E_c^2}{R}\left[J_1^2(\beta) + J_2^2(\beta) + \ldots\right] \\ &= \frac{E_c^2}{R}\left[J_0^2(\beta) + 2(J_1^2(\beta) + J_2^2(\beta) + \ldots)\right] \\ &= P_c\left[J_0^2(\beta) + 2(J_1^2(\beta) + J_2^2(\beta) + \ldots)\right]. \end{aligned} \tag{5.1.26}$$

Note that the unmodulated carrier power is $P_c = E_c^2/R$. A property of the Bessel functions is that the sum $J_0^2(\beta) + 2[J_1^2(\beta) + J_2^2(\beta) + \ldots] = 1$, and so the total average power is equal to the unmodulated carrier power. This is in fact the expected result, as the amplitude of the carrier does not change by frequency modulation. In effect, when modulation is applied, the total power that was originally in the carrier is redistributed between all components of the spectrum.

Example 5.4 A 20 W unmodulated carrier is frequency-modulated with a sinusoidal signal such that the peak frequency deviation is 8 kHz. The frequency of the modulating signal is 1 kHz. Calculate the average power output by summing the powers for all the sideband components.

Solution

The total average power output is 20 W, unmodulated or modulated. To check that this is also the value obtained from the sum of the squares of the Bessel function coefficients, we have

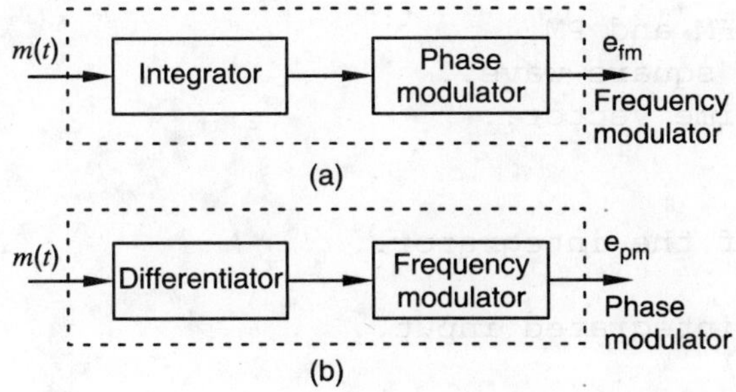

Figure 5.3 *Frequency and phase modulation*

$$\beta = \frac{\delta}{f_{\mathrm{m}}} = 8/1 = 8.$$

The Bessel function values for $\beta = 8$ are obtained using the MATLAB function *bessel* $j(n, 8)$ for $n = 0$ to 12. Substituting the values in equation (5.1.26), we get

$$\begin{aligned} P_{\mathrm{T}} &= 20(0.17^2 + 2[0.23^2 + 0.11^2 + 0.29^2 + 0.1^2 + 0.19^2 + 0.34^2 + 0.32^2 \\ &\quad +0.22^2 + 0.13^2 + 0.06^2 + 0.03^2 + 0.01^2]) = 20(0.9951) = 19.902W \approx 20 \text{ W} \end{aligned}$$

5.2 Phase Modulation

Phase modulation is closely related to FM. It is possible to obtain FM from PM by the *Armstrong system*. PM is not used in practical analog modulation systems.

If the phase ϕ in the equation $e = A \sin(\omega_c t + \phi)$ is varied so that its magnitude is proportional to the instantaneous amplitude of the modulating voltage, the resulting wave is phase-modulated. The expression for the PM wave is

$$e = A \sin(\omega_c t + \phi_m \sin \omega_m t), \tag{5.2.27}$$

where ϕ_m is the maximum value of phase change introduced by this particular modulating signal and is proportional to the maximum amplitude of this modulation. Now we can rewrite (5.2.27) as

$$e = A \sin(\omega_c t + m_p \sin \omega_m t), \tag{5.2.28}$$

where $m_p = \phi_m$ is the modulation index for phase modulation.

It may be noted that FM and PM are inseparable, and one can be generated from the other. This is evident from Figure 5.3, which shows the compatibility between the two systems. To compare the shapes of waveforms obtained for FM and PM for a periodic modulating signal represented by $e_{mp} = 3$ square $(2\pi\ f_m\ t)$, we take up the following MATLAB simulation.

Let the period of the above square wave be 1ms. The following MATLAB script file generate the PM and FM waves and plot them.

```
% MATLAB script to generate FM and PM..
% The modulating signal is a square wave..
t=linspace(0,.002,1000); % time vector..
%%
fm=1000;kp=3;
tau=.0001; % time constant of the integrator..
emp=3*square(2*pi*fm*t); %
emf(1)=0; % first sample of integrated input..
kf=kp;
for l=2:length(emp)
    emf(l)=trapz(t(1:l),emp(1:l))/tau; % Trapezoidal
integration of end; % input (modulating) signal..
epm=8*sin(40000*pi*t+kp*emp);
efm=8*sin(40000*pi*t+kf*emf);
subplot(411),plot(t,emp,'r'), axis([0 .002 -10 10]);
xlabel('t'), ylabel('e_{mp}'),title('Modulating Signal for
PM/FM'); subplot(412),plot(t,emf);
xlabel('t'), ylabel('e_{mf}'),title('Integrated Modulating
Signal'); subplot(413),plot(t,epm,'r');
xlabel('t'), ylabel('e_{pm}'),title('PM_{out}for Square wave
input'); subplot(414),plot(t,efm,'b');hold off;
xlabel('t'), ylabel('e_{fm}'),title('FM_{out}for
Square wave input');
%%End of pmafm.m
```

The resulting PM and FM waveforms are given in Figure 5.4.

Now, we can try to simulate the PM and FM output for a triangular input signal. Note that this is the signal that is actually applied to PM system, as per Figure 5.3 in the preceeding case) to generate FM.

```
%% MATLAB  script file to generate and plot PM
% and  FM wave for a triangular input signal..
%
em=4*linspace(-1,1,1000); % One period of triangular wave
emp=em'*ones(1,10);
emp=emp(:)';
t=linspace(0,.0001,length(emp));% Time vector for 10 cycles of
%% input wave.
fc=10^5;
kf=2000*pi;%FM modulation index.
kp=pi/2;%PM modulation index.
tau=kp/(2*pi*kf); % Time constant of the integrator..
epm=8*sin(2*pi*fc*t+kp*emp);%PM out generated..
emf(1)=0;
for p=2:length(emp)
   emf(p)=trapz(t(1:p),emp(1:p))/tau; % Trapezoidal integration of
end; % input signal...
```

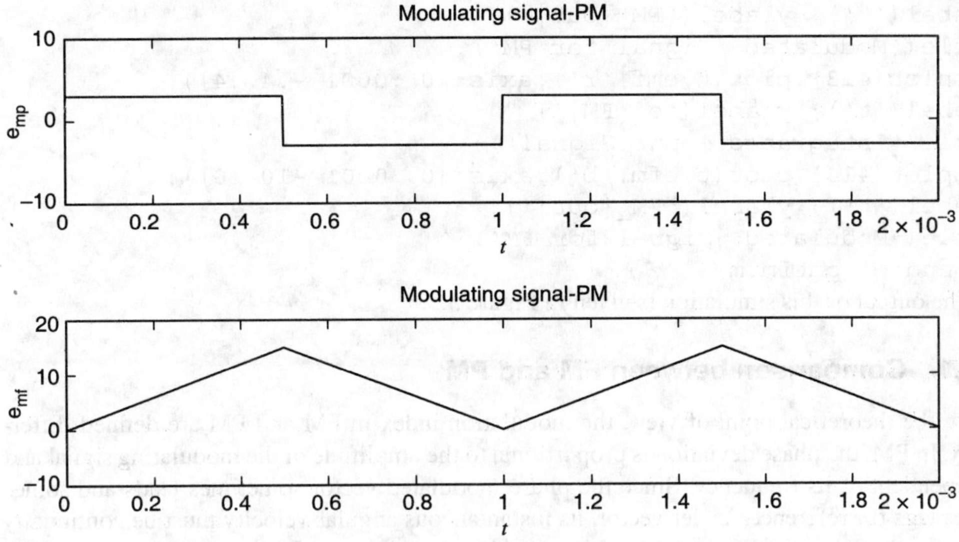

Figure 5.4 *PM and FM waveforms for a square wave input*

```
efm=8*sin(2*pi*fc*t+kp*emf);% FM out generated.
subplot(411),plot(t,emp,'r'), axis([0 .0001 -5 5]);
xlabel('t'); ylabel('e_{PM/FM}');
title('Modulating Signal for FM/PM');
subplot(412),plot(t,epm,'b'), axis([0 .0001 -10 10]);
xlabel('t'); ylabel('PM_{out}');
title('Modulated  Signal for PM');
subplot(413),plot(t,emf,'r'),axis([0 .0001 -.4 .4]);
xlabel('t'); ylabel('e_{FM}');
title('Integrated Input Signal');
subplot(414),plot(t,efm,'b'),axis([0 .0001 -10 10]);
xlabel('t'); ylabel('FM_{out}');
title('Modulated  Signal for FM');
%% end of pmtfm.m
```

The output of this simulation is given in Figure 5.5.

5.2.1 Comparison between FM and PM

From the theoretical point of view, the modulation index in FM and PM are defined differently. In PM, the phase deviation is proportional to the amplitude of the modulating signal and independent of its frequency. Since the phase modulated vector sometimes leads and sometimes lags the reference carrier vector, its instantaneous angular velocity must be continually changing between the limits imposed by ϕ_m; thus, some form of frequency change must be taking place. In FM, the frequency deviation is proportional to the amplitude of the modulating voltage. Moreover, if we take a reference vector rotating with a constant angular velocity corresponding to the carrier frequency, then the FM vector will have a phase lead or lag with respect to the reference as its frequency oscillates between $f_c - \delta$ and $f_c + \delta$. Hence, FM must be a form of PM. In FM, the larger the frequency deviation, the larger is the phase deviation, and so the latter depends at least to a certain extent on the amplitude of the modulation, just as in PM. But the modulation index of FM is also *inversely proportional to the modulating frequency.* This means that under identical conditions, *FM and PM are indistinguishable for a single modulating frequency.* When the modulating frequency is changed, the PM modulation index remains the same while as the FM modulation index increases as the modulation frequency is reduced, and vice versa. A numerical example will clearly illustrate this point.

Example 5.5 A 30 MHz carrier is modulated by a 800 Hz audio sine wave. If the peak carrier voltage is 5V and the maximum deviation is 16 kHz, write the equation of this modulated wave for: (a) FM and (b) PM. If the modulating frequency is now changed to 3.2 kHz everything else remaining the same, write a new equation for: (c) FM and (d) PM.

Solution
Calculating the angular frequencies, we have

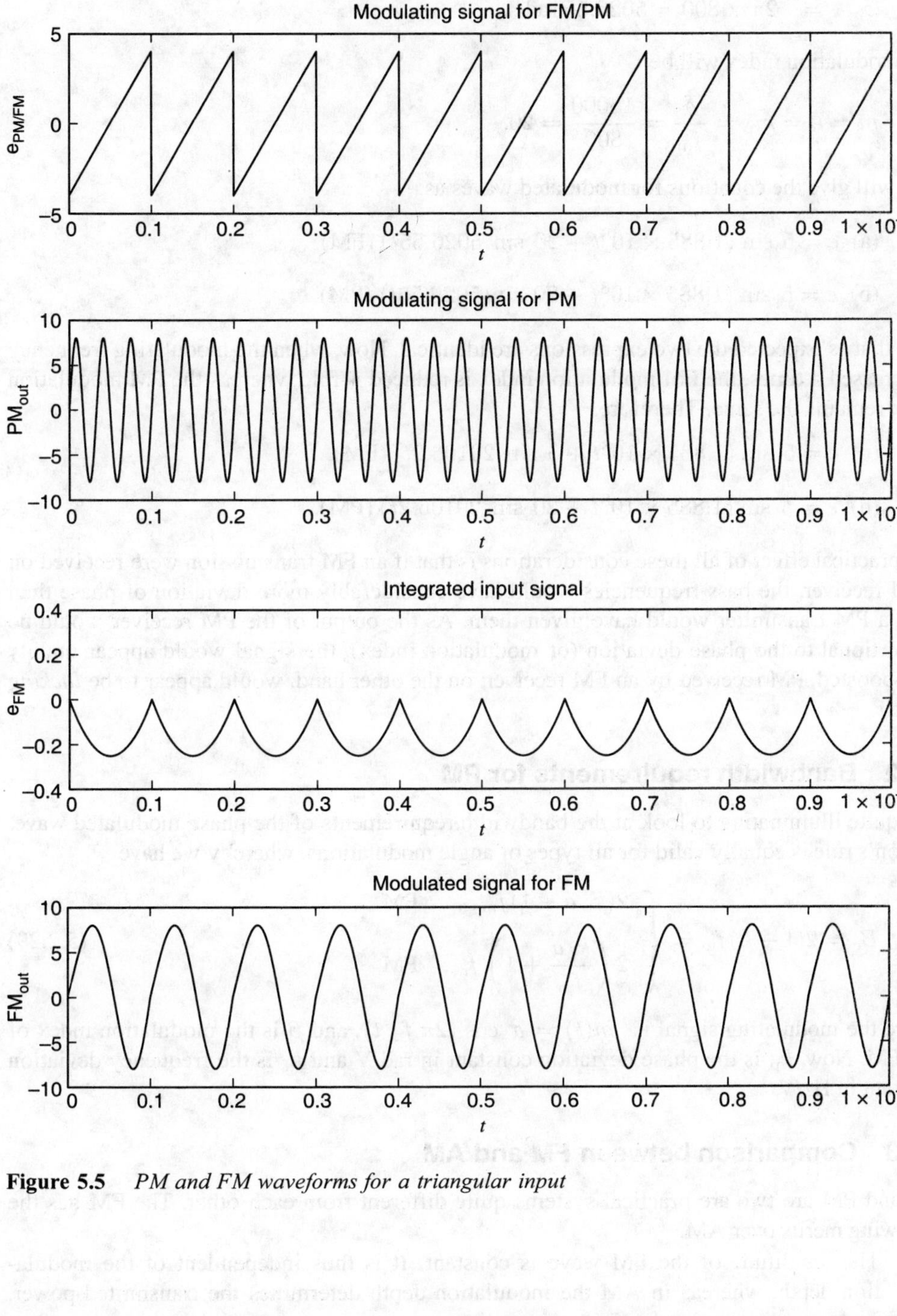

Figure 5.5 *PM and FM waveforms for a triangular input*

$$\omega_c = 2\pi \times 30 \times 10^6 = 1.885 \times 10^8 \text{ rad/s.}$$

$$\omega_m = 2\pi \times 800 = 5026.55 \text{ rad/s.}$$

The modulation index will be

$$m = \beta = m_p = \frac{\delta}{f_m} = \frac{16000}{800} = 20.$$

This will give the equations for modulated waves as:

(a) $e = 5 \sin(1.885 \times 10^8 t + 20 \sin 5026.55t)$ (FM).

(b) $e = 5 \sin(1.885 \times 10^8 t + 20 \sin 5026.55t)$ (PM).

Note that as expected the two expressions are identical. Now, when the modulating frequency is increased 4 times, the FM modulation index is reduced 4-fold whereas the PM modulation index remains the same. Therefore,

(c) $e = 5 \sin(1.885 \times 10^8 t + 5 \sin 20106.2t)$ (FM).

(d) $e = 5 \sin(1.885 \times 10^8 t + 20 \sin 20106.2t)$ (PM).

The practical effect of all these considerations is that if an FM transmission were received on a PM receiver, the bass frequencies would have considerably more deviation of phase than what a PM transmitter would have given them. As the output of the PM receiver would be proportional to the phase deviation (or modulation index), the signal would appear unduly bass-boosted. PM received by an FM receiver, on the other hand, would appear to be *lacking in bass.*

5.2.2 Bandwidth requirements for PM

It is quite illuminating to look at the bandwidth requirements of the phase modulated wave. Carson's rule is equally valid for all types of angle modulations, whereby we have

$$B_c = 2(1+\beta)f_m = \begin{cases} 2(k_p\, a + 1)f_m, & \text{PM} \\ 2\left(\dfrac{k_f\, a}{f_m} + 1\right) f_m, & \text{FM} \end{cases} \tag{5.2.29}$$

where the modulating signal is $m(t) = a \cos(2\pi f_m t)$, and β is the modulation index of PM/FM. Now, k_p is the phase deviation constant in rad/V and k_f is the frequency deviation constant in Hz/V.

5.2.3 Comparison between FM and AM

FM and PM are two are practical systems quite different from each other. The FM has the following merits over AM.

- The amplitude of the FM wave is constant. It is thus independent of the modulation depth, whereas in AM the modulation depth determines the transmitted power.

Practically, this means that in FM transmitters low-level modulation may be used but all subsequent amplifiers can be class C and therefore more efficient. Moreover, all the transmitted power in FM is useful, whereas in AM most of it is in the transmitted carrier which contains no useful information.

- FM receivers can be fitted with amplitude limiters to remove the amplitude variations caused by noise, and this makes FM reception much more immune to noise than AM.
- It is possible to reduce noise still further by increasing the deviation. AM does not have this feature. Increasing modulation index beyond 100% causes severe distortion in AM.
- The guard band allocation by ITU for FM stations provide good adjacent channel rejection.
- FM broadcast band is in the upper VHF and UHF frequency ranges which happen to be less noisy than the MF and HF frequency ranges occupied by AM broadcast.
- At the FM broadcast frequencies *space wave propagation* is used and so the radius of operation is limited to slightly more than *line of sight*. It is therefore possible to operate several independent transmitters on the same frequency with considerably less interference than would be possible with AM.

The following are some of the disadvantages of FM transmission.

- A much wider channel is required for FM, up to 10 times as large as that required for AM. This is the most significant disadvantage of FM.
- FM transmitting and receiving equipments tend to be more complex, especially for modulation and demodulation.
- As the FM reception is limited to line of sight, the area of reception for FM is much smaller than for AM. This is a disadvantage for FM mobile communications over a wide area.

5.3 FM Transmission and Noise

Frequency modulation is much more immune to noise than AM, and is significantly more immune than PM. In FM, the signal-to-noise ratio (SNR) is improved as the modulating frequency is reduced. Under identical conditions, FM will be 4.75 dB better in SNR than PM. It can be shown that in FM the noise has a higher effect on the higher modulating frequencies than the lower ones.

5.3.1 Pre-emphasis and de-emphasis

It can be shown that in FM, if the high frequencies are boosted before modulation (they are subsequently attenuated after demodulation at the receiver to null the effect), an improvement in SNR results. This boosting of high frequencies in accordance with a prearranged curve is called *pre-emphasis*. The corresponding compensation done at the receiver is called *de-emphasis*. Typical circuits for achieving the same are shown in Figure 5.6. The amount of pre-emphasis for FM broadcast in US is 75 μs, whereas the European and Australian TV transmissions use 50 μs. Note that the pre-emphasis circuit is a *high-pass filter* and the de-emphasis circuit is, therefore, a *lowpass filter.*

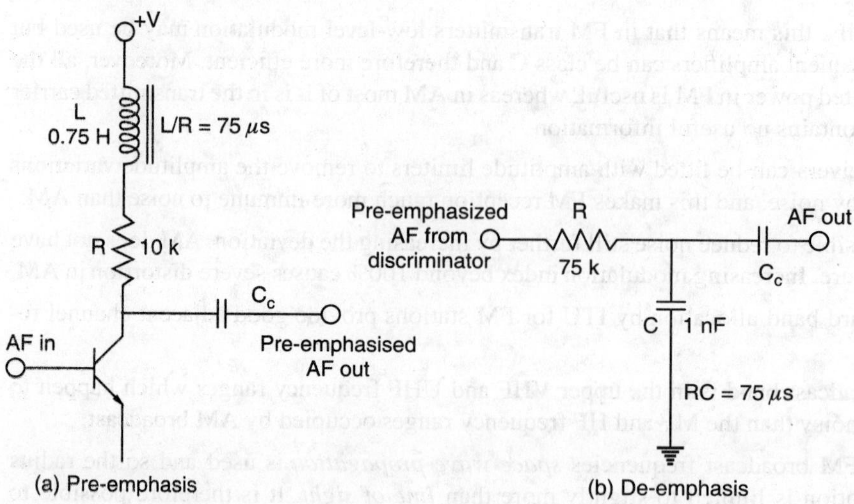

Figure 5.6 *Pre-emphasis and de-emphasis circuits*

5.4 Frequency Modulator Circuits

One method of generation of FM is apparent. If either the capacitance or inductance of an LC oscillator tank is varied, FM will result. In such a system a voltage variable inductance (capacitance) is placed across the tank, and the tank is tuned so that in the absence of modulation, the oscillator frequency is equal to the desired carrier frequency. The inductance (capacitance) of the variable element is changed with the modulating voltage, increasing (or decreasing) as the modulating voltage increases positively and going the other way when modulation becomes negative. The greater the departure of the modulating voltage, the greater is the reactance variation and hence the frequency variation. When the modulating voltage is zero, the variable reactance will have its average value. There are several two-terminal (example, the *varactor diode*) and three-terminal devices (FETs, BJTs, and vacuum tubes) whose reactance can be varied by the application of voltage.

5.4.1 Varactor diode modulator

In a varactor diode, the junction capacitance under reverse-bias condition varies linearly with the applied voltage. Figure 5.7 shows a reactance modulator using varactor diode. It is seen from the figure that the diode is reverse biased to provide the junction capacitance effect, and as the bias is varied by modulating voltage in series to the voltage $-V_b$, the junction capacitance will vary, causing the oscillator frequency to change accordingly. This is the simplest reactance modulator and is often used for automatic frequency control (AFC) and remote tuning.

5.4.2 Basic reactance modulator using FET

With reference to Figure 5.8(a), we make the following assumptions in analysing the operation of a reactance modulator using FET.

- The bias network current i_b is negligible compared to the drain current.

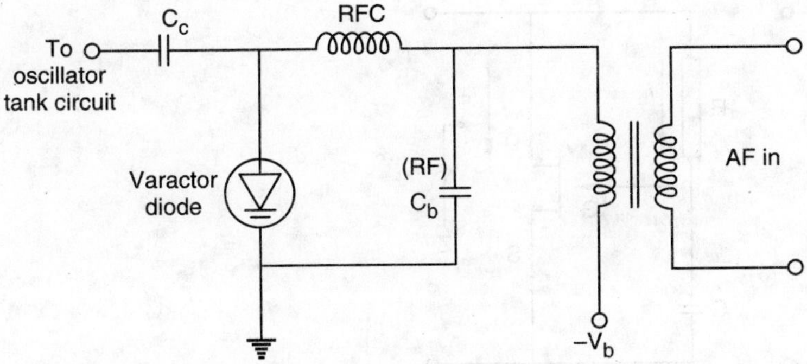

Figure 5.7 *Varactor diode frequency modulator*

- The drain to gate impedance (X_c) must be very much larger than the gate–source impedance (R in this case).

There are three other versions of FET reactance modulators, as depicted in Figure 5.8 (b) to (d).

Analysis of FET reactance modulator: Case 1

To determine z, a voltage v_{ds} is applied between the drain and source and the resulting current i_d is calculated. Dividing the voltage v_{ds} by the current i_d, we will get the impedance between the drain and the source. Now, we have

$$v_g = i_b\, R = \frac{R\, v_{ds}}{R - j\, X_c}. \tag{5.4.30}$$

The FET drain current i_d is given by

$$i_d = g_m\, v_g = \frac{g_m\, R\, v_{ds}}{R - j\, X_c}. \tag{5.4.31}$$

Therefore, the impedance seen at the terminals D and S is

$$z = \frac{v_{ds}}{i_d} = \frac{v_{ds}}{\frac{g_m\, R\, v_{ds}}{R - j\, X_c}} = \frac{R - j\, X_c}{g_m\, R} = \frac{1}{g_m}\left(1 - \frac{j\, X_c}{R}\right). \tag{5.4.32}$$

Since $X_c \gg R$, the preceding equation will reduce to

$$z = -j\, \frac{X_c}{g_m\, R}. \tag{5.4.33}$$

This impedance is a capacitive reactance, which can be written as

$$X_{eq} = \frac{X_c}{g_m\, R} = \frac{1}{2\pi\, f\, g_m\, RC} = \frac{1}{2\pi\, f\, C_{eq}}. \tag{5.4.34}$$

Thus under conditions stipulated here, the output impedance of the FET is a pure capacitive reactance and is given by

$$C_{eq} = g_m\, RC. \tag{5.4.35}$$

We can make the following observations from the preceding equation.

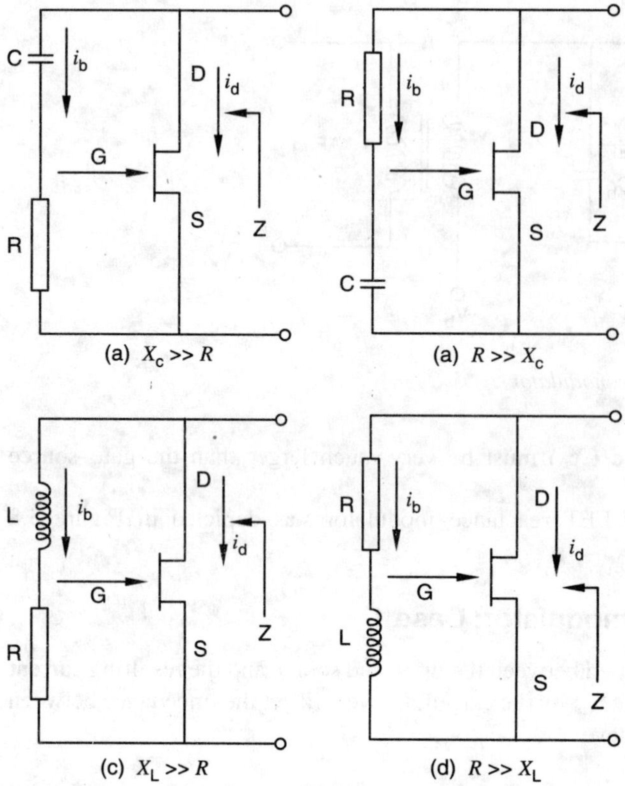

Figure 5.8 *Basic reactance modulators using FET*

1. The equivalent capacitance depends on the device transconductance (g_m) and can be varied with bias voltage.
2. The capacitance can be originally adjusted to any value, within limits, by varying the components R and C.
3. The expression $g_m\, RC$ has the correct dimension of capacitance as the dimensions of R and g_m cancel each other out.
4. If the gate-to-source resistance is not small compared with the gate–drain impedance, z would have a resistive component as well.

In practice, the gate–drain impedance is made five to ten times the gate–source impedance. Let $X_c = n\, R$ (at the carrier frequency). Then, we have

$$X_c = \frac{1}{\omega\, C} = n\, R$$

$$C = \frac{1}{\omega\, nR} = \frac{1}{2\pi\, f\, nR}. \tag{5.4.36}$$

Substituting (5.4.36) in (5.4.35),

Table 5.1 *Summary of various reactance modulators*

Name or Type of Circuit	Z_{gd}	Z_{gs}	Condition	Reactance relationship
RC capacitive	C	R	$X_C \gg R$	$C_{eq} = g_m RC$
RC inductive	R	C	$R \gg X_C$	$L_{eq} = \dfrac{RC}{g_m}$
RL inductive	L	R	$X_L \gg R$	$L_{eq} = \dfrac{L}{g_m R}$
RL capacitive	R	L	$R \gg X_L$	$C_{eq} = \dfrac{g_m L}{R}$

$$C_{eq} = g_m RC = \frac{g_m R}{2\pi f nR}$$

$$C_{eq} = \frac{g_m}{2\pi f n}. \tag{5.4.37}$$

Equation (5.4.37) is a very useful formula to find C_{eq} when f, n, and g_m are given.

Analysis of FET reactance modulator: Case 2

In this case, as referred in Figure 5.8 (b), we interchange R and C and assume that $R >> X_c$. Everything else remaining the same, we have,

$$v_g = i_b.\frac{1}{j\omega C} = \frac{v_{ds}}{R + \frac{1}{j\omega C}}.\frac{1}{j\omega C} = \frac{v_{ds}}{1 + j\omega RC}$$

$$i_d = g_m v_g = \frac{g_m.v_{ds}}{1 + j\omega RC}$$

$$z = \frac{v_{ds}}{i_d} = \frac{1 + j\omega RC}{g_m} = \frac{R}{g_m}(\frac{1}{R} + j\omega C) \approx \frac{j\omega RC}{g_m}.$$

The expression for z shows that it is inductive, and the equivalent inductance is given by $L_{eq} = \frac{RC}{g_m}$. Now, the other two cases of FET reactance modulators are shown in Figure 5.8 (c) and (d). The analysis of output impedance as seen across drain and source is left as an exercise to the reader. The four types of reactance modulators that will provide useful results are summarised in Table 5.1.

Example 5.6 Obtain the value of capacitive reactance got from a reactance FET whose g_m is 20 millisiemens. Assume that the gate–source resistance is one-tenth of the gate–drain capacitive reactance and that the carrier frequency is 8 MHz.

Solution

$$C_{eq} = \frac{g_m}{2\pi f n} \quad \text{Therefore, } 2\pi f C_{eq} = \frac{g_m}{n} = \frac{1}{X_{C_{eq}}}.$$

$$X_{C_{eq}} = \frac{n}{g_m} = \frac{10}{20 \times 10^{-3}} = 500\Omega.$$

Example 5.7 The mutual conductance of an FET varies linearly with applied gate voltage between the limits of 0 and 10 mS. The FET is used as a capacitive reactance modulator with $X_{C_{gd}} = 10\ R_{gs}$, and is placed across an oscillator circuit tuned to 80 MHz by a 60 pF fixed capacitor. What will be the total frequency variation when the transconductance of the FET is varied from 0 to 10 mS, by the modulating voltage?

Solution

Let us assume:

C_n = minimum equivalent capacitance of the reactance FET

C_m = maximum equivalent capacitance of the reactance FET

f_n = minimum frequency

f_m = maximum frequency

f = average frequency

δ = maximum deviation.

Then,

$$C_n = 0.$$

$$C_m = \frac{g_m}{2\pi f n} = \frac{10 \times 10^{-3}}{2\pi \times 8 \times 10^7 \times 10} = \frac{10 \times 10^{-11}}{16\pi}.$$

$$= 1.989 \times 10^{-12} = 1.989 pF.$$

$$\frac{f_m}{f_n} = \frac{1/2\pi\sqrt{LC}}{1/2\pi\sqrt{L(C+C_m)}} = \sqrt{\frac{C+C_m}{C}} = \sqrt{1+\frac{C_m}{C}} = \sqrt{1+\frac{1.989}{60}}$$

$$= \sqrt{1.03316} = 1.0164.$$

Now,

$$\frac{f_m}{f_n} = \frac{f+\delta}{f-\delta}.$$

$$f + \delta = (f - \delta) \times 1.0164 = 1.01644 f - 1.01644\,\delta.$$

$$2.01644\,\delta = 0.01644f$$

$$\Rightarrow \delta = \frac{0.01644f}{2.01644} = 0.01644 \times 80 \times 10^6/2.01644 = 0.6522\,\text{MHz}.$$

The total frequency variation is $2\delta = 2\times\ 0.6522$ MHz $= 1.3045$ MHz.

Example 5.8 It is required to provide a maximum deviation of 75 kHz for the 88 MHz carrier frequency of an FM transmitter. A FET is used as a capacitive reactance modulator, and the linear region of its $g_{\mathrm{m}} - v_{\mathrm{gs}}$ curve lies from 360 μS (at which $V_{\mathrm{gs}} = -4$ V) to 880 μS (at which $V_{\mathrm{gs}} = -0.6$ V). Assuming that R_{gs} is one-twelfth of $X_{c_{gd}}$, calculate: (a) the rms value of the required modulating voltage and (b) the value of the fixed capacitance and inductance of the LC oscillator tuned circuit across which the reactance modulator is connected.

Solution

(a) V_{m} (peak-to-peak) $= 4 - 0.6 = 3.4$ V.

V_{m} (rms) $= 3.4/2\sqrt{(2)} = 1.202$ V

(b) $C_{\mathrm{n}} = \dfrac{g_{\mathrm{m,min}}}{2\pi\ f\ n} = \dfrac{3.6 \times 10^{-4}}{2\pi \times 8.8 \times 10^7 \times 12} = 0.0542$ pF.

$$C_{\mathrm{m}} = \frac{C_{\mathrm{n}}\,g_{\mathrm{m,max}}}{g_{\mathrm{m,min}}} = 0.0542 \times \frac{880}{360} = 0.1326\,\text{pF}.$$

Now we have

$$\frac{f_{\mathrm{m}}}{f_{\mathrm{n}}} = \frac{1}{2\pi\sqrt{L(C + C_{\mathrm{n}})}} \div \frac{1}{2\pi\sqrt{L(C + C_{\mathrm{m}})}} = \sqrt{\frac{C + C_{\mathrm{m}}}{C + C_{\mathrm{n}}}}$$

$$\Rightarrow \left(\frac{f_{\mathrm{m}}}{f_{\mathrm{n}}}\right)^2 = \frac{C + C_{\mathrm{m}}}{C + C_{\mathrm{n}}}\ \frac{f_{\mathrm{m}}^2}{f_{\mathrm{n}}^2} - 1 = \frac{C + C_{\mathrm{m}}}{C + C_{\mathrm{n}}} - 1$$

$$\Rightarrow \frac{f_{\mathrm{m}}^2 - f_{\mathrm{n}}^2}{f_{\mathrm{n}}^2} = \frac{C + C_{\mathrm{m}} - C - C_{\mathrm{n}}}{C + C_{\mathrm{n}}}$$

$$\Rightarrow \frac{(f_{\mathrm{m}} + f_{\mathrm{n}})(f_{\mathrm{m}} - f_{\mathrm{n}})}{f_{\mathrm{n}}^2} = \frac{4\ f\ \delta}{f_{\mathrm{n}}^2} \approx \frac{4\ f\ \delta}{f^2} = \frac{4\ \delta}{f} = \frac{C_{\mathrm{m}} - C_{\mathrm{n}}}{C + C_{\mathrm{n}}}$$

$$\Rightarrow C + C_{\mathrm{n}} = \frac{(C_{\mathrm{m}} - C_{\mathrm{n}})f}{4\ \delta}$$

$$\Rightarrow C = \frac{(C_{\mathrm{m}} - C_{\mathrm{n}})f}{4\delta} - C_{\mathrm{n}}$$

$$= \frac{(0.1326 - 0.0542) \times 88}{4 \times 0.075} - 0.0542$$

$$\approx \frac{0.0784 \times 88}{0.3} = 23\text{pF}$$

$$f = \frac{1}{2\pi\sqrt{L(C + C_{\text{av}})}} \approx \frac{1}{2\pi\sqrt{LC}}$$

$$L = \frac{1}{4\pi^2 f^2 C} = \frac{1}{4\pi^2 \times 8.8^2 \times 10^{14} \times 2.3 \times 10^{-11}}$$

$$= \frac{10^{-3}}{39.47 \times 77.44 \times 2.3} = \frac{10^{-5}}{70.32} = 1.422 \times 10^{-7}$$

$$= 0.1422\ \mu\text{H}$$

5.5 FM Transmitters

Usually, direct FM at the final carrier is not feasible because of the problem of maintaining high-frequency stability of the carrier while at the same time obtaining adequate frequency deviation. Hence, a subcarrier is frequency modulated and then a combination of a *rrequency multiplier* and a mixer is used to obtain FM with the final carrier frequency.

With frequency multiplication, the instantaneous frequency is multiplied by a factor, say n. Hence, if $f_{\text{i}} = f_{\text{O}} + \Delta f$, then the signal passing through a frequency multiplier will make it $nf_{\text{i}} = nf_{\text{O}} + n\Delta f$. Frequency multiplication can be achieved by passing the signal through a class C amplifier and tuning the output to the desired harmonic.

With frequency mixing, the deviation is not altered. For example, if a signal with instantaneous frequency $f_{\text{i}} = f_{\text{O}} + \Delta f$ is passed through a mixer that is also fed with a local oscillator output signal with frequency f_{LO}, then the mixer output circuit can be tuned to the difference frequency $f_{\text{O}} + \Delta f - f_{\text{LO}} = f_{\text{IF}} + \Delta f$.

5.5.1 Frequency-stabilised FM transmitter

An LC oscillator may be directly frequency-modulated to produce relatively large deviation, but it may be difficult to maintain the stability of the unmodulated frequency. A crystal oscillator cannot be used along with a reactance modulator directly even though the stability provided by it is necessary in commercial FM transmitters as per FCC recommendations. The required stability can be achieved by using *automatic frequency control*. Figure 5.9 shows one such arrangement. A sample of the output signal, taken from the final driver stage, is mixed with a signal from a stable crystal oscillator. The difference frequency output from the mixer is fed into the discriminator circuit, where it is demodulated. The demodulated output consists of the original modulating signal and any variations caused by the drifting of the nominal frequency of the LC oscillator. A lowpass filter removes the signal component while leaving the drift component, which is fed back as a control bias to the oscillator. This is connected in such a way as to shift the nominal frequency in the desired direction to reduce the drift.

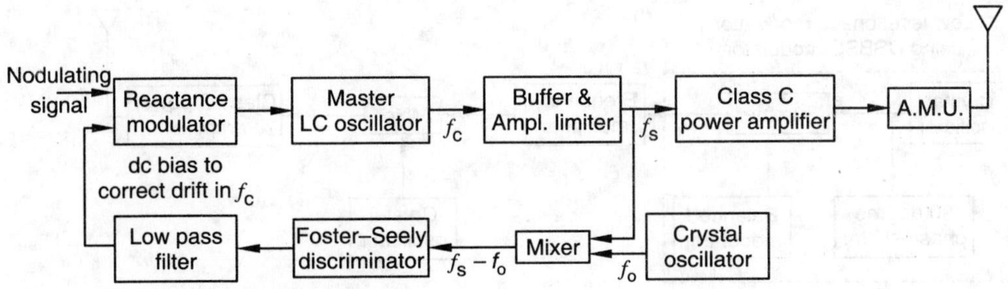

Figure 5.9 *FM transmitter with AFC*

5.5.2 Indirect FM transmitter using PM

In this technique, a highly stabilised FM transmitter is build upon a crystal oscillator that is the carrier source for a phase modulator. The modulating signal (to the FM) is integrated before it is fed to this PM and hence it can produce FM in effect. The FM output obtained from the phase modulator is frequency multiplied, mixed (for obtaining the required frequency deviation) and fed to the class C power amplifier before it is transmitted.

5.5.3 Armstrong's method for FM

In Armstrong's method, the initial modulation takes place as an amplitude modulated DSBSC signal so that a crystal controlled oscillator can be used for high stability of carrier frequency. To see the connection between these two, we reproduce the equation for a low-level phase modulated signal as:

$$e(t) = V_c \cos(\omega_c t + K_p v_m(t)). \tag{5.5.38}$$

Expanding the preceding equation using trigonometric identities,

$$e(t) = V_c[\cos \omega_c t. \cos(K_p v_m(t)) - \sin \omega_c t. \sin(K_p v_m(t))]. \tag{5.5.39}$$

Now the peak phase deviation in the angle $K_p\ v_m(t)$ can be kept small such that cos $(K_p\ v_m(t)) \approx 1$ and sin $(K_p\ v_m(t)) \approx K_p\ v_m(t)$. Then the expansion for the PM wave becomes

$$e(t) = V_c[\cos \omega_c t - K_p v_m(t) \sin \omega_c t]. \tag{5.5.40}$$

The first term of the expansion is the unmodulated carrier and the second term is a DSBSC term with carrier shifted by 90°. Thus, by summing the DSBSC signal with a carrier term, phase modulation can be achieved through AM.

Figure 5.10 shows how an Armstrong's modulator works. The crystal oscillator generates a subcarrier which can be as low as 100 kHz. One output from the oscillator is phase-shifted by 90° to produce the sine term in (5.5.40), which is balance modulated with $v_m(t)$ to produce DSBSC. This is combined with the direct output from the oscillator in the summing amplifier producing a low-level PM output. Now, as the modulating signal is passed through an integrator before it is phase modulated, at the output of the summing amplifier we get a low-level FM. The succeeding stages raises the frequency and deviation of the FM signal to the final level.

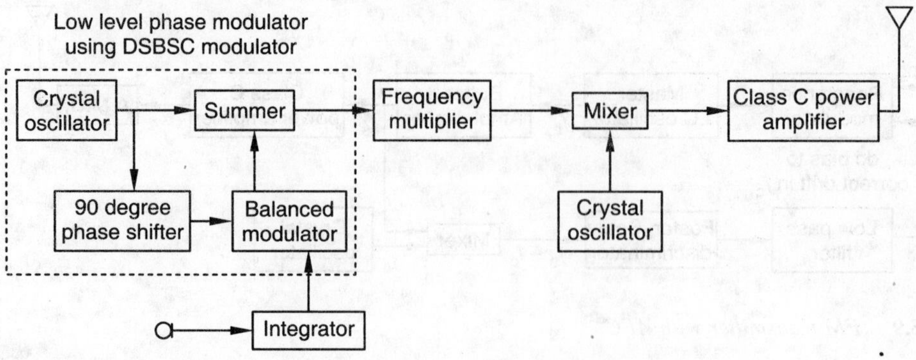

Figure 5.10 *Armstrong's FM transmitter*

5.6 FM Broadcast

The major requirement of FM broadcast is excellent fidelity, as music is the main program material. As will be made clear in a subsequent section, FM audio broadcast is superior to AM audio on many counts. Due to this reason, FM is the modulation used for audio signals in TV.

The baseband for FM audio broadcast extends from 50 Hz to 15 kHz, with a maximum allowable deviation of $\pm$ 75 kHz. Channel spacing is 200 kHz and the power outputs can be as high as 100kW.

Recently, FM stereo is catching up in broadcasting. Both two-channel and four-channel stereos are in vogue. Moreover, some FM stations frequency-division multiplex an additional background music (BGM) channel on their carrier for use in public buildings. This is licensed as a subsidiary communications authorisation (SCM).

The block diagram of a two-channel FM stereo broadcast transmitter (which can be received by a monophonic FM receiver) is given in Figure 5.11. In that, the left and right audio channels are mixed to provide two signals, one of which is a balanced *monophonic* signal. The first is the sum of two input channels and the second is the difference. The sum channel

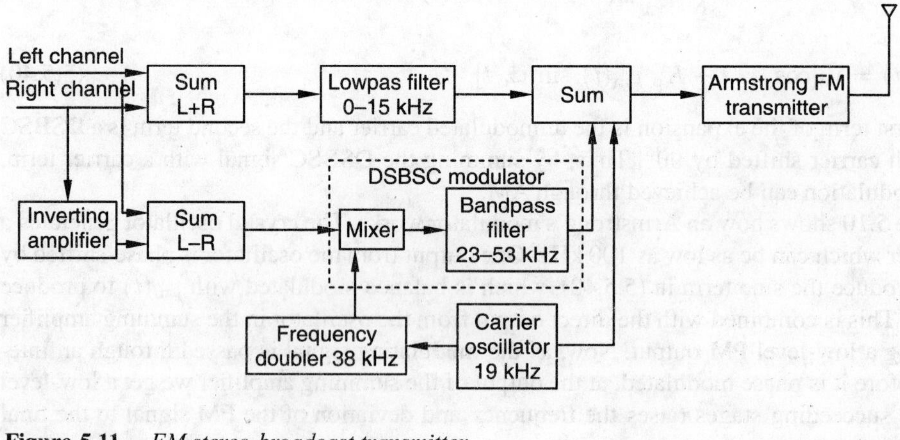

Figure 5.11 *FM stereo broadcast transmitter*

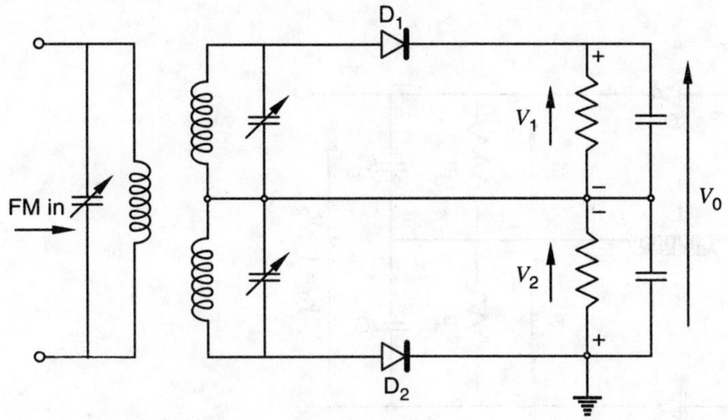

Figure 5.12 *Balanced slope detector*

is modulated in the baseband between 50 and 15000 Hz. The difference signal is DSBSC-modulated in the 23 to 53 kHz slot around a carrier at 38 kHz. A pilot carrier at 19 kHz is also transmitted. The sum signal can be demodulated by a monophonic FM receiver to provide normal FM monophonic reception. A stereo FM receiver can also retrieve the difference signal and combine the two to produce the original left and right channels.

5.7 FM Detectors

The FM detector circuit should have the basic property of getting an output voltage that varies linearly with the frequency of the input signal. The *slope detector* is the basic form of such a circuit, even though the linearity of its response is not good.

5.7.1 Round Travis detector

This is also known as *balanced slope detector*. It consists of two slope detectors, one tuned to f_{01} above the incoming carrier frequency and the other to f_{02} below the carrier frequency. The envelope detectors combine to give a differential output. The output of the detector is $V_O = |V_1| + |V_2|$ and will have a S shape when plotted against frequency. When the incoming signal is unmodulated, the output is balanced to zero. When the carrier deviates toward f_{01}, $|V_1|$ increases and $|V_2|$ decreases, and hence the output is positive. On the other hand, when the carrier deviates towards f_{02}, $|V_1|$ decreases and $|V_2|$ increases, and therefore, the output goes negative. Figure 5.12 shows the circuit diagram of a balanced slope detector.

5.7.2 Foster–Seeley discriminator

This discriminator is also known as the *centre tuned discriminator*. Figure 5.13 shows the circuit diagram of a Foster–Seeley discriminator. The Foster–Seeley discriminator is derived from the Round–Travis detector and is widely used in practical FM receivers. In this, both the primary and secondary are tuned to the carrier frequency. This greatly simplifies the alignment and yields better linearity compared within the balanced slope detector. It can be shown that

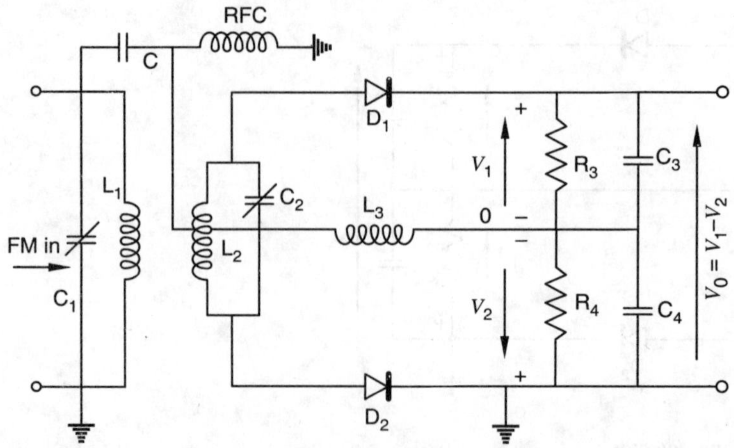

Figure 5.13 *Foster–Seeley discriminator*

the voltage applied to each diode is the sum of the primary voltage and the corresponding half secondary-voltage. It will also be shown that the primary and secondary voltages are:

- Exactly 90° out of phase when the input frequency is f_c.
- Less than 90° out of phase when f_{in} is higher than f_c.
- More than 90° out of phase when f_{in} is below f_c.

The result of this situation is that individual output voltages will be equal only at f_c. At all other frequencies, the output of one diode will be greater than that of the other. Which diode has the larger output will depend entirely on whether f_{in} is above or below f_c, as the required magnitude of the output will depend on the deviation of the input frequency from f_c.

5.7.3 Ratio detector

In the Foster–Seeley discriminator, amplitude changes in the input voltage will give rise to amplitude changes in the resulting output voltage. This necessitates the use of an amplitude limiter in the prior stages. A modified limiter circuit, known as the *ratio detector*, dispenses with this requirement. It is illustrated in Figure 5.14. The ratio detector is also widely used in practical receivers. Compared with the Foster–Seeley discriminator, its linearity is a notch poorer.

5.7.4 PLL-based FM detection

The block diagram of a phase-locked-loop (PLL)-based FM demodulator is shown in Figure 5.15. The phase detector, which is basically a balanced modulator, produces an average (or low-frequency) output voltage that is a linear function of the phase difference between the two input signals. The low-frequency component is selected by the lowpass filter that also removes much of the noise. The filtered signal is amplified through amplifier A and is passed as a control voltage to the voltage-controlled oscillator (VCO), where, it results in frequency modulation of the VCO free-running frequency, f_o. When the loop is in lock, the VCO frequency follows or tracks the incoming frequency. For example, when the instantaneous incoming frequency increases, the control voltage will cause the VCO frequency to increase.

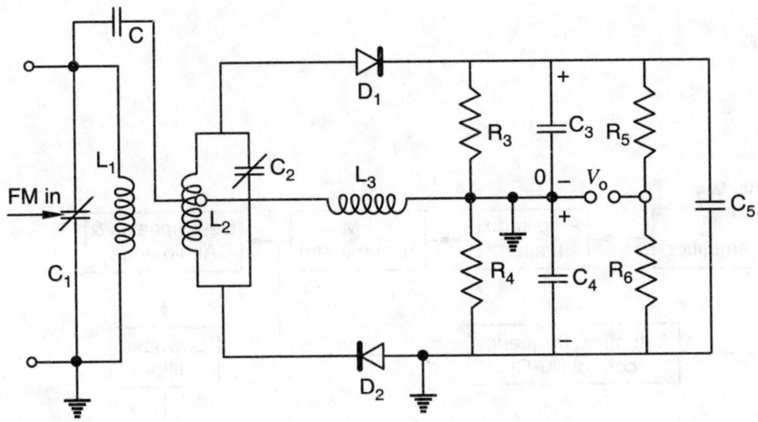

Figure 5.14 *Ratio detector*

The instantaneous frequency of the FM wave is $f_i = f_c + ke_m(t)$. For the VCO, the instantaneous frequency can be written as $f_{vco} = f_o + k_{vco}V_c(t)$, where f_o is the free-running frequency of the VCO, and k_{vco} is the frequency deviation constant of the VCO. For the VCO frequency to track the instantaneous incoming frequency:

$$f_{vco} = f_i \tag{5.7.41}$$

or, $\quad f_o + k_{vco}V_c(t) = f_c + ke_m(t).$

From this it can be shown that

$$V_c(t) = \frac{f_c - f_o + ke_m(t)}{k_{vco}}. \tag{5.7.42}$$

The VCO can be tuned so that $f_o = f_c$, and hence the control voltage is equal to a constant times the modulating signal voltage. Thus, the output of the amplifier under lock condition, $v_o(t)$, which is the same as $V_c(t)$, becomes the demodulated output signal.

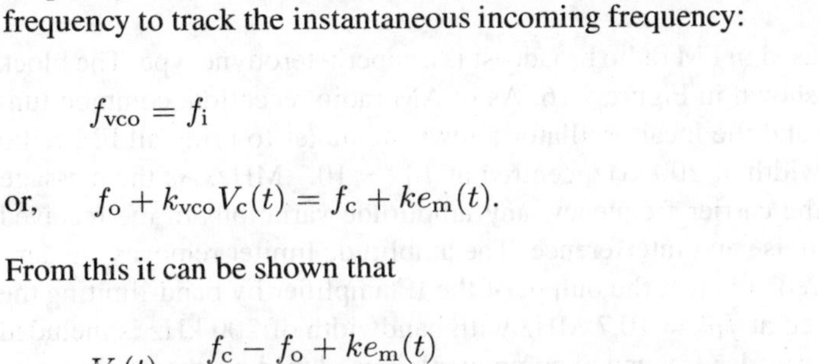

Figure 5.15 *PLL-based FM demodulator*

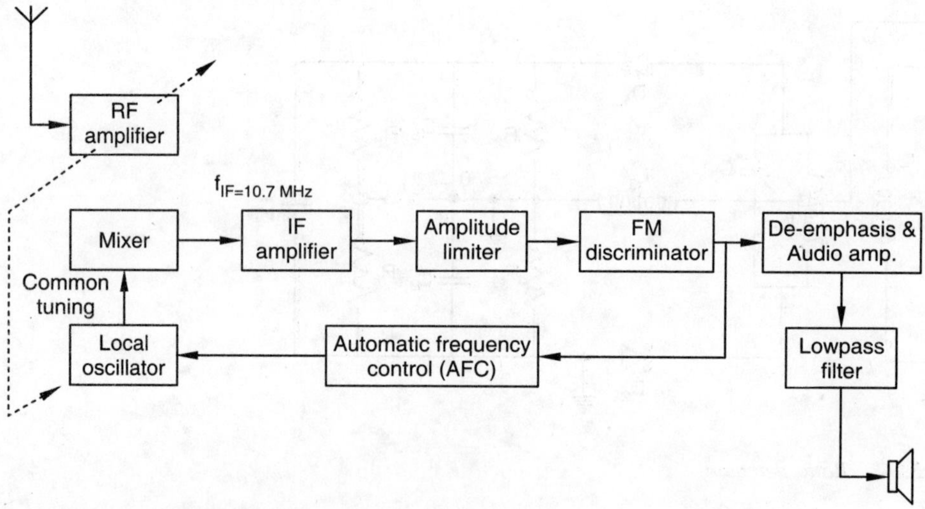

Figure 5.16 *Superheterodyne FM receiver*

5.8 FM Radio Receiver

The receiver most commonly used in FM radio broadcast is a superheterodyne type. The block diagram of such a receiver is shown in Figure 5.16. As in AM radio reception, common tuning between the RF amplifier and the local oscillator allows the mixer to bring all FM radio signals to a common IF bandwidth of 200 kHz, centred at $f_{\text{IF}} = 10.7$ MHz. As the message signal $m(t)$ is embedded in the carrier frequency, any amplitude variations in the received signal are a result of additive noise and interference. The amplitude limiter removes any amplitude variations in the received signal at the output of the IF amplifier by band-limiting the signal. A bandpass filter centred at $f_{\text{IF}} = 10.7$ MHz with bandwidth of 200 kHz is included in the limiter to remove higher-order frequency components introduced by the nonlinearity inherent in the *hard limiter.*

In FM superheterodyne reception, the demodulator is a *limited frequency discriminator.* The message is transmitted by variations of the instantaneous frequency of a sinusoidal carrier wave while maintaining the amplitude constant. Hence any variation of carrier amplitude at the receiver input is due to noise or interference. The use of amplitude limiter after the IF stage removes amplitude variations by clipping the modulated waves at the IF output. The resulting square wave is rounded off by a bandpass filter that suppresses the harmonics of the carrier frequency and hence the resultant output of the bandpass filter is sinusoidal.

A balanced frequency discriminator is used for frequency demodulation. The resulting message signal is then passed to the audio amplifier, which performs the dual functions of de-emphasis and amplification. The output of the audio amplifier is further lowpass-filtered to remove out-of-band noise, and its output is used to drive a loud speaker.

5.9 Recent Trends

We will now consider a new development in radio transmission and reception: software defined radio (SDR). It is one of the most advanced technology evolved in recent times.

5.9.1 Software-defined radio (SDR)

An SDR system is a radio communication system that uses software for the modulation and demodulation of radio signals. An SDR performs significant amounts of signal processing in a general-purpose computer, or a reconfigurable piece of digital electronics. The goal of this design is to produce a radio that can receive and transmit a new form of radio protocol just by running a new software.

Software radios have significant utility for military and cell phone services, both of which must serve a wide variety of changing radio protocols in real time. The hardware of an SDP typically consists of a superheterodyne RF front end that converts RF signals from and to analog IF signals, and analog-to-digital converters and digital-to-analog converters that are used to convert a digitised IF signal to analog form.

SDR can currently be used to implement simple radio modem technologies. In the long run, SDR is expected by its proponents to become the dominant technology in radio communications. *GNURadio* is a project to implement software-defined radio as free software.[3]

> **FCC Definition of Software-Defined Radio**
>
> *Software-defined radio (SDR)*: A radio that includes a transmitter in which the operating parameters of frequency range, modulation type or maximum output power (either radiated or conducted), or the circumstances under which the transmitter operates in accordance with Commission rules, can be altered by making a change in software without making any changes to hardware components that affect the radio frequency emissions.

Ideal software-defined radio

The ideal Software-defined radio (SDR) receiver scheme would be to attach an analog-to-digital converter to an antenna. A computer would read the converter, and then the software would transform the stream of data from the converter to any other form.

An ideal transmitter would be similar. A computer program would generate a stream of numbers. These would be sent to a digital-to-analog converter connected to a radio antenna. Note that the ideal scheme is not practical.

The system of systems construct

To maintain interoperability, the SDR adapts a system of systems construct. This is illustrated in Figure 5.17.

Benefits of SDR

Apart from interoperability, the following benefits can be identified for the SDR technology.

[3] See *https://sourceforge.net/projects/opensdr*
http://www.sdrforum.org for details.

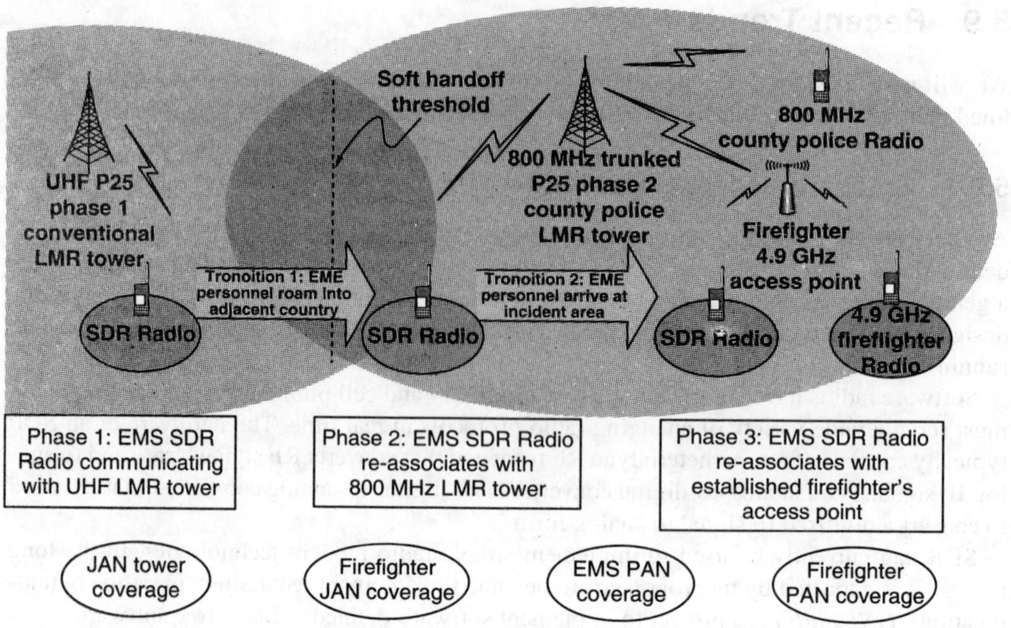

Figure 5.17 *The system of systems construct of SDR*

- Reduction of costs associated with upgrading/modifying equipment.
- Ability to adapt to evolving technologies and standards: SDR technology can significantly facilitate the introduction of new technologies and evolving standards to the field by simplifying the process required to take advantage of technologies or become compliant with new standards.
- Ease of operation: SDR technology can be used to hide implementation details from users to provide more seamless radio operation to end users.
- Performance optimisation: the capability to adjust the parameters of the waveform as required to most efficiently interface with the rest of the system, given the varying atmospheric, interference and other conditions, could significantly improve the quality of radio system performance especially under adverse conditions. One specific aspect of this concept involves capabilities to utilise the spectrum in a more efficient manner (spectrum efficiency through dynamic spectrum allocation).

Exercises

1. What is the highest audio frequency allowed for use in commercial FM radio broadcasting? (Answer: 15 kHz)
2. What will be the deviation, when the percentage of modulation in FM is 80%? (Answer: $0.8 \times \pm$ 75 kHz = $\pm$ 60kHz)
3. What is the minimum distance in kilometres between two class C FM stations operating on the same frequency? (Answer: 289.67 km)

4. An FM broadcast carrier is deviated by $\pm$ 30 kHz. Find the percentage of modulation. (Answer: $\frac{30\ \text{kHz}\times 100}{75\ \text{kHz}} = 40\%$)
5. A carrier wave has an unmodulated frequency of 4 MHz. Sketch the instantaneous frequency time curve when it is frequency modulated by a 1000 Hz sawtooth waveform which varied 10 V peak-to-peak about a zero DC value. The frequency deviation constant is 5 kHz per volt. Use MATLAB for coding.
6. A carrier is frequency-modulated by a sinusoidal signal of 10 V peak and frequency 4 kHz, the frequency deviation constant being 2 kHz/V. Calculate: (a) the peak frequency deviation and (b) the modulation index. (Answer: (a) 20 kHz (b) 5)
7. An FM carrier is limited to a maximum deviation of 75 kHz. Compare the bandwidths when the modulating signal is sinusoidal of frequency of (a) 100 Hz (b) 1 kHz and (c) 10 kHz, the carrier reaching maximum deviation in each case. State how you would determine the bandwidth limits. (Answer: (a) 150.2 kHz, (b) 152 kHz, (c) 170 kHz, in each case Carson's rule is applied, as $\beta > 6.0$)
8. A 2 kHz square wave is used to frequency modulate a carrier, producing a peak deviation of 75kHz. Assuming that the square wave harmonics up to and including the 11th need to be taken into account, calculate the deviation ratio and, using Carson's rule find the bandwidth. (Answer: $\frac{75}{2\times 11} = 3.41$, 2(75 + 22)=194 kHz)
9. The phase deviation constant in a phase modulation system is 0.02 rad/V. Calculate the peak phase deviation when a sinusoidal modulating signal of 15 V peak is applied to it. Does the peak phase deviation depend on the modulating frequency? (Answer: 0.3rad, No)
10. A carrier is frequency modulated by a ramp voltage that starts at zero and rises to 6 V in 3 ms when it abruptly drops to zero. Given that the frequency deviation constant is 4 kHz/V and the unmodulated sinusoidal carrier frequency is 200 kHz, plot the modulated angle over time base of 10 ms. Use MATLAB for coding.
11. A ramp voltage is used to phase modulate a 40 kHz carrier. The ramp rises linearly from zero to 5 V in 250 μ s, when it abruptly drops to zero. Calculate the instantaneous frequency over the modulating period. The phase deviation constant is $\pi\ rad/V$.(Answer: $80\pi \times 10^6 + 40\pi \times 10^3$)
12. A sinusoidal signal of 20V peak is used to phase modulate a carrier, the phase deviation constant being 1rad/V. Determine (a) the peak phase deviation and (b) the equivalent frequency deviation for a modulating frequency of 10 kHz. Repeat the calculations for the modulating frequency increased to 20 kHz, all other things remaining the same. (Answer: (a) 20 rad (b) 200 kHz; 20 rad and 400 kHz)
13. A 4000 Hz sinusoidal carrier signal of 10 V peak is used to phase modulate a carrier, the phase deviation constant being 2 rad/V. Obtain the practical bandwidth required for the modulated signal. (Answer: 2(4 + 20) kHz)
14. A binary sequence 101010 is used to phase modulate a carrier, the bit period being 4 times the carrier period. Given that the modulated carrier amplitude is 1 V peak, draw accurately the modulated waveform showing the full modulating sequence.
15. A 100 kHz carrier is frequency modulated to produce a peak deviation of 1500 Hz. This FM is passed through a 4 $\times$ 4 $\times$ 5 frequency multiplier chain, the output of which is mixed with an oscillator signal and the difference frequency taken as the new output.

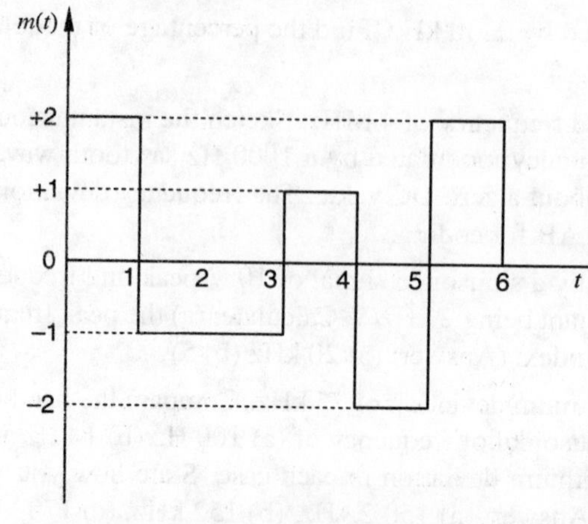

Figure 5.18 *Quadrature level signal*

Determine the frequency of the oscillator required to produce a 100 kHz FM output and the peak deviation of the output. (Answer: 7.9 MHz, 120 kHz)

16. An Armstrong FM modulator uses a primary oscillator of 1 MHz, and the maximum phase deviation for good linearity is limited to 10^c rad with sinusoidal modulation. What is the corresponding peak frequency deviation at modulating frequency of 100 Hz? If the transmitted signal is to have a maximum peak frequency deviation of 30 kHz at a carrier frequency of 120 MHz, specify the frequency multiplication factor needed, and suggest a chain of doublers and triplers to accomplish this. For the preceding chain, specify the local oscillator frequency needed for the final frequency conversion. (Answer: 100×10^c Hz, 144, $2^4 \times 3^2$, 24 MHz)
17. An FM transmitter has a carrier oscillator with a rest frequency of 3.5 MHz. The oscillator shifts the frequency by $\pm$ 1.6 kHz when a 3.6 V peak-to-peak message signal is applied. The frequency multiplier section following the oscillator has three frequency triplers. Find the transmitted carrier rest frequency, the deviation, and the percentage modulation at the antenna. (Answer: $3.5 \times 27 = 94.5$ MHz, $\pm$ 1.6 kHz $\times\ 27 = \pm$ 43.2 kHz, $\frac{43.2\times100}{75} = 57.6\%$)
18. The message signal $m(t)$ shown in Figure 5.18 is given into an FM modulator with peak frequency deviation $f_d = 25$ Hz/V. Plot the frequency deviation in hertz and the phase deviation in radians.
19. An angle modulated signal is given by:

$$u(t) = 100 \cos\,[2\pi\, f_c\, t + 4 \sin\, 2\pi\, f_m\, t]$$

where $f_c = 10$ MHz and $f_m = 1000$ Hz.

(a) Assuming that this is an FM signal, determine the modulation index and the transmitted signal bandwidth.
(b) Repeat part (a) if f_m is doubled.
(c) Assuming that this is a PM signal, determine the modulation index and the transmitted signal bandwidth.

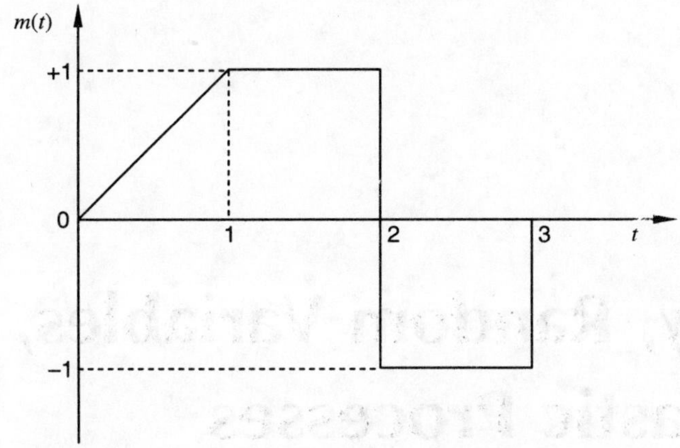

Figure 5.19 *Multilevel signal*

(d) Repeat part (c) if f_m is doubled.
(Answer: (a) 4, 10 kHz (b) 2, 12 kHz (c) 4, 10 kHz (d) 4, 20 kHz)

20. A carrier oscillator at 5.05 MHz is deviated $\pm$ 3.125 kHz by a 2.85 V peak-to-peak message signal. Determine the radiated carrier frequency, the deviation, and the percentage modulation at the antenna, and the message voltage needed to cause 100% modulation when the multiplier section has one doubler and two triplers. (Answer: 90.9 MHz, $\pm$ 56.25 kHz, $\frac{56.75\times 100}{75} = 75\%$, $\frac{75\times 2.85}{56.25} = 3.8$ V peak-to-peak.)

21. Signal $m(t)$ shown in Figure 5.19 is used to frequency modulate/phase modulate a carrier.
    (a) Find a relation between k_p and k_f such that the maximum phase of the modulated signals in both cases are equal.
    (b) If $k_p = f_d = 1$, what is the maximum instantaneous frequency in each case?

22. Determine the amplitude and phase of various frequency components of a PM signal with $k_p = 1$ and $m(t)$ a periodic signal given by

$$m(t) = \begin{cases} +1, & 0 \le t \le \frac{T_m}{2} \\ -1, & \frac{T_m}{2} \le t \le T_m. \end{cases}$$

for one period.

Chapter 6

Probability, Random Variables, and Stochastic Processes

This chapter is devoted to the study of random processes and their properties. Random processes provide a good model for both information sources and noise. Stochastic processes is a synonym for random processes. Random processes are analyzed using statistical and probabilistic methods. If the process is stationary, a Fourier transform can simplify a random looking function into a collection of coefficients, thus transforming the process to the frequency domain. *A random process does not have a predominant amplitude or frequency.* Random processes occur in nature.

When a signal is transmitted through a communication channel, there are two types of imperfections that cause the received signal to be different from the transmitted signal. One class of imperfections is deterministic in nature, such as linear and nonlinear distortion and inter symbol interference. The second class is nondeterministic, such as addition of noise and multipath fading.

6.1 Probability Space

The most fundamental concept in any probabilistic model is the concept of *random experiment*, which is any experiment whose outcome, for some reason, cannot be predicted with certainty. Flipping a coin, rolling a die and drawing a card from a deck are examples of random experiments. What is common in all these cases is that the result (or outcome) of the experiment is uncertain. A random experiment has certain *outcomes* that are elementary results of the experiment. In flipping a coin, 'head' and 'tail' are the possible outcomes. The set of all possible outcomes is called the *sample space* and is denoted by Ω. Outcomes are denoted by ω's and certainly, each ω lies in Ω, i.e., $\omega \in \Omega$.

6.1.1 Discrete sample space

A sample space is *discrete* if the number of its elements are finite or countably infinite. Or else it is a *nondiscrete* sample space. All the random experiments given here have a discrete sample space. If one chooses randomly a number between 0 and 1, then the random space corresponding to this random experiment is non-discrete.

6.1.2 Events

Events are subsets of the sample space for which a probability can be defined. For discrete sample spaces, any subset of the sample space is an event, that is, a probability can be defined for it. For example, in the tossing of a coin, events like 'getting a head' or 'getting a tail' can be considered. For a non-discrete sample space, not every subset of Ω can be assigned a probability without sacrificing the basic intuitive properties of probability. To overcome this difficulty, we define a σ-field $\mathcal{B}$ on the sample space as a collection of subsets of Ω such that the following conditions are satisfied.

1. $\Omega \in \mathcal{B}$.
2. If the subset (event) $E \in \mathcal{B}$, then $E^c \in \mathcal{B}$, where E^c denotes the complement of E.
3. If $E_i \in \mathcal{B}$ for all i, then $\cup_{i=1}^{\infty} E_i \in \mathcal{B}$.

We define a *probability measure* *P* on $\mathcal{B}$ as a set function assigning non-negative values of all events E in $\mathcal{B}$ such that the following conditions are satisfied:

1. $0 \leq P(E) \leq 1$ for all $E \in \mathcal{B}$.
2. $P(\Omega) = 1$.
3. For disjoint events $E_1, E_2, E_3, \ldots$ (i.e., events for which $E_i \cap E_j = \phi$ for all $i \neq j$, where ϕ is the null set), we have $P(\cup_{i=1}^{\infty} E_i) = \sum_{i=1}^{\infty} P(E_i)$.

The triple $(\Omega, \mathcal{B}, P)$ is called a *probability space*.

Some of the important properties of probability are listed below:

1. $P(E^c) = 1 - P(E)$: (The probability of 'non-occurrence of an event' is one minus the probability of its occurrence).
2. $P(\phi) = 0$: (probability of a 'non-occurrent event', i.e., null event is zero).
3. $P(E_1 \cup E_2) = P(E_1) + P(E_2) - P(E_1 \cap E_2)$.
4. If $E_1 \subset E_2$, then $P(E_1) \leq P(E_2)$.

6.1.3 Mutually exclusive events

Two events A and B are said to *mutually exclusive or disjoint* if their corresponding sets have no common elements, that is, if

$$A \cap B = \phi. \tag{6.1.1}$$

Several sets $A_1, A_2 \ldots$ are mutually exclusive if

$$A_i \cap A_j = \phi \text{ for every } i \text{ and } j \neq i. \tag{6.1.2}$$

6.1.4 Conditional probability

Let us assume that the two events E_1 and E_2 are defined on the same probability space with corresponding probabilities $P(E_1)$ and $P(E_2)$. Then if an observer receives the information that the event E_2 has in fact occurred, the observer's probability about the event E_1 will not be $P(E_1)$ any more. In fact, the information that the observer received changes the probabilities of various events, and new probabilities, called *conditional probabilities*, are defined.[1]

[1] Note: The terms *a priori* and *a posteriori* are often used for the probabilities $P(A_i)$ and $P(A_i|B)$.

The conditional probability of the event E_1 given that event E_2 has occurred is defined by

$$P(E_1|E_2) = \begin{cases} \dfrac{P(E_1 \cap E_2)}{P(E_2)}, & P(E_2) \neq 0 \\ 0, & \text{otherwise.} \end{cases} \tag{6.1.3}$$

6.1.5 Statistical independence

If it happens that $P(E_1|E_2) = P(E_1)$, then the knowledge of E_2 does not change the probability of occurrence of E_1. In this case, the events E_1 and E_2 are said to be *statistically independent*. For statistically independent events, $P(E_1 \cap E_2) = P(E_1).P(E_2)$.

Example 6.1 In rolling a die, the probability of

A= {the outcome is even} is $P(A) = P(2) + P(4) + P(6) = \frac{1}{2}$.

The probability of

B= {the outcome is less than 4} is $P(B) = P(3) + P(2) + P(1) = \frac{1}{2}$.

In this case, $P(B|A) = \frac{P(B \cap A)}{P(A)} = \frac{P(2)}{1/2} = \frac{1}{3}$. Note that $P(B|A) = \frac{1}{3} \neq P(B)$. Hence, the two events are *not statistically independent.*

Example 6.2 In a coin tossing experiment, the sample space consists of the outcomes h and t:

$$S = \{h, t\}.$$

We consider a new experiment of tossing the coin three times. The possible outcomes are:

$$S = \{hhh, hht, hth, htt, thh, tht, tth, ttt\}.$$

We shall assume that all elementary events have the same probability (fair coin). In this case, the probability of the elementary event is $\frac{1}{8}$. The event

$$\{\text{Heads at the first two tosses}\} = \{hhh, hht\},$$

consists of two outcomes, hhh and hht; hence its probability is $\frac{2}{8}$.

Example 6.3 A biased coin is tossed till a head appears for the first time. What is the probability that the number of required tosses is odd?

Solution

Let

$$\begin{aligned} A_i &= \{\text{Head appears at the } i\text{th toss for the first time}\} \\ &= \{\underbrace{T, T, T, \ldots, T}_{i-1}, H\}. \end{aligned}$$

Assuming that each trial is independent of the rest,

$$P(A_i) = P(\{T, T, T, \ldots, T, H\}) = P(T)P(T)\ldots P(T)P(H) = q^{i-1}.p,$$

where $P(H) = p,\ \ P(T) = q = 1 - p$. Then

$$P\,(\text{'heads appears on an odd toss'}) = P(A_1 \cup A_3 \cup A_5 \cup \ldots)$$

$$= \sum_{i=0}^{\infty} P(A_{2i+1}) = \sum_{i=0}^{\infty} q^{2i}p = p\sum_{i=0}^{\infty} q^{2i}$$

$$= \frac{p}{1-q^2} = \frac{p}{(1-q)(1+q)}$$

$$= \frac{1}{1+q} = \frac{1}{2-p}.$$

6.1.6 Bayes' theorem

If the events $\{E_i\}_{i=1}^{n}$ make a partition of the sample space Ω, i.e., if the following two conditions

$$\cup_{i=1}^{n} E_i = \Omega \tag{6.1.4}$$

$$E_i \cap E_j = \phi, \text{ for all } 1 \leq i, j \leq n \text{ and } i \neq j. \tag{6.1.5}$$

are satisfied, then, for an event A we have the conditional probabilities $\{P(A|E_i)\}_{i=1}^{n}$, $P(A)$ can be obtained by applying the *total probability theorem* stated as

$$P(A) = \sum_{i=1}^{n} P(E_i)P(A|E_i). \tag{6.1.6}$$

The Bayes' theorem gives the conditional probabilities $P(E_i|A)$ by the following relation.

$$P(E_i|A) = \frac{P(E_i).P(A|E_i)}{\sum_{j=1}^{n} P(E_j)P(A|E_j)}. \tag{6.1.7}$$

6.2 Random Variables

A *real random variable* is a mapping from the sample space Ω to the set of real numbers. A random variable (χ) is a function such that, for every x, both real and complex, there exists a probability of $\chi = x$. A schematic diagram representing a random variable is given in Figure 6.1.

Random variables are denoted by capital letters A, B, χ, etc., and individual values of the random variable χ are $\chi(\omega)$. A random variable is *discrete* if the range of its values is finite or countably infinite. This range is usually denoted by $\{x_i\}$. In other words, a discrete random variable is a variable whose sample space consists of integer values $n \in 1, 2, 3, \ldots$

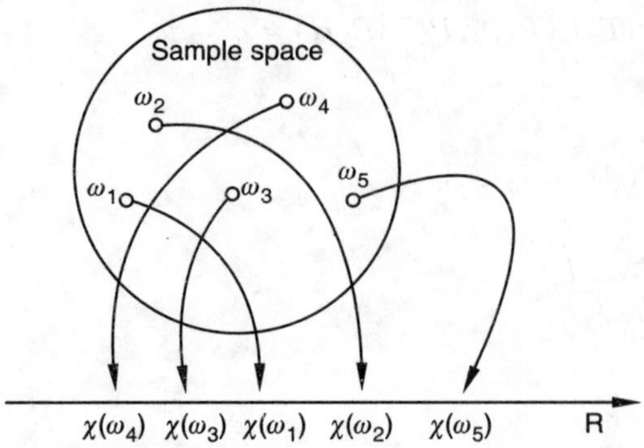

Figure 6.1 *Random variables as a mapping from Ω to $\mathcal{R}$*

6.2.1 Cumulative distribution function

The *cumulative distribution function* (CDF) of a random variable X is defined as

$$F_X(x) = P(\omega \in \Omega; X(\omega) \leq x). \tag{6.2.8}$$

which can be simply written as

$$F_X(x) = P(X \leq x) \tag{6.2.9}$$

The CDF has the following properties:

1. $0 \leq F_X(x) \leq 1$.
2. $F_X(x)$ is non-decreasing.
3. $\lim_{x \to -\infty} F_X(x) = 0$ and $\lim_{x \to +\infty} F_X(x) = 1$.
4. $F_X(x)$ is continuous from the right, i.e., $\lim_{\epsilon \downarrow 0} F_X(x + \epsilon) = F_X(x)$.
5. $P(a < X \leq b) = F_X(b) - F_X(a)$.
6. $P(X = a) = F_X(a) - F_X(a^-)$.

For discrete random variables, $F_X(x)$ is a staircase function. A random variable is called *continuous* if $F_X(x)$ is a continuous function. A random variable is called *mixed* if it is neither discrete nor continuous.

6.2.2 Probability density function

The *probability density function* (PDF) of a random variable X is defined as the derivative of $F_X(x)$; i.e.,

$$f_X(x) = \frac{d}{dx} F_X(x). \tag{6.2.10}$$

In case of discrete or mixed random variables, the probability density function involves impulses. The basic properties of probability density function are listed below.

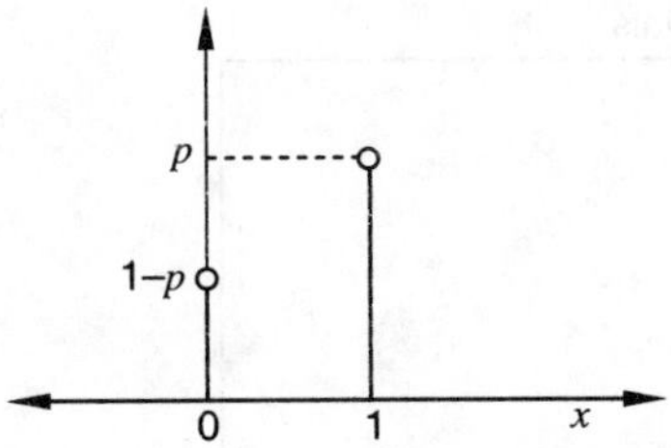

Figure 6.2 *The PMF for Bernoulli random variable*

1. $f_X(x) \geq 0$.
2. $\int_{-\infty}^{\infty} f_X(x)dx = 1$.
3. $\int_{a^+}^{b^+} f_X(x)dx = P(a < X \leq b)$.
4. In general, $P(X \in A) = \int_A f_X(x)dx$.
5. $F_X(x) = \int_{-\infty}^{x^+} f_X(u)du$.

For discrete random variables, we define *probability mass functions* (PMF) , which is defined as $\{p_i\}$, where $p_i = P(X = x_i)$. Obviously, for all i we have $p_i \geq 0$ and $\sum_i p_i = 1$.

6.2.3 Important random variables

There are several commonly used random variables in communications systems, as described below.

Bernoulli random variable

This a random variable taking two values, 1 and 0, with probabilities p and $1-p$. A Bernoulli random variable is good model for a binary data generator. Also, it can be used to model channel errors. The PMF for the Bernoulli random variable is illustrated in Figure 6.2.

Binomial random variable

This is a discrete random variable giving the number of 1s in a sequence of n independent Bernoulli trials. The PMF is given by

$$P(X = k) = \begin{cases} \binom{n}{k} p^k(1-p)^{n-k}, & 0 \leq k \leq n \\ 0, & \text{otherwise.} \end{cases} \tag{6.2.11}$$

This random variable models, for example, the total number of bits received in error when a sequence of n bits is transmitted over a channel with bit-error probability p. The PMF for

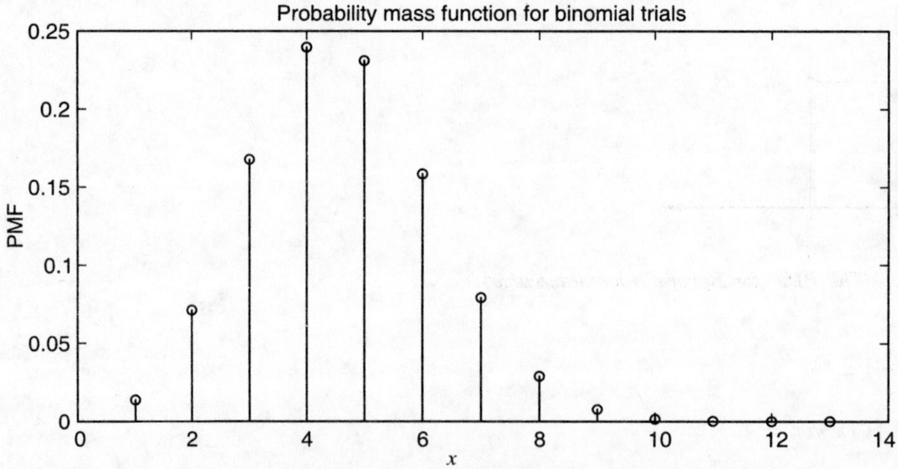

Figure 6.3 *The PMF for binomial random variable*

binomial random variable is given in Figure 6.3. The MATLAB script m-file to generate the same is also given below.

```
% M-file to plot the Binomial Random Variable..
clf;
clear all;
n=12;
p=.3;
for k=0:n
  P(k+1)=nchoosek(n,k)*p^k*(1-p)^(n-k);
end;
stem(P), xlabel('x'), ylabel('PMF');
title('Probability Mass Function for Binomial Trials');
```

Example 6.4 Coin A has a probability of $\frac{1}{4}$ of getting a head and a probability of $\frac{3}{4}$ of getting a tail, coin B is a fair coin. Each coin is tossed 4 times. Let X denote the random variable of number of heads resulting from tossing coin A, and let Y denote the resulting number of heads from coin B.

1. What is the probability that $X = Y = 2$?
2. What is the probability that $X = Y$?
3. What is the probability that $X > Y$?
4. What is the probability that $X + Y \leq 5$?

Solution

1. Note that $P\{X = Y = 2\} = P\{X = 2\}.P\{Y = 2\}$. Since the processes are binomially distributed, their PMF are $\binom{n}{k}p^k(1-p)^{n-k}$. Hence,

$$P\{X=2\} = \binom{4}{2}\left(\frac{1}{4}\right)^2\left(\frac{3}{4}\right)^2 = 6\left(\frac{9}{256}\right) = \frac{27}{128}$$

$$P\{Y=2\} = \binom{4}{2}\left(\frac{1}{2}\right)^2\left(\frac{1}{2}\right)^2 = 6\left(\frac{1}{16}\right) = \frac{3}{8}.$$

Therefore the required probability is:

$$P\{X=Y=2\} = \frac{81}{1024}.$$

2. In this case the required probability is given by

$$P\{X=Y\} = \sum_{i=0}^{4} P\{X=Y=i\}.$$

So we have

$$\begin{aligned} P\{X=Y\} &= \binom{4}{0}\left(\frac{3}{4}\right)^4\binom{4}{0}\left(\frac{1}{2}\right)^4 + \binom{4}{1}\left(\frac{1}{4}\right)\left(\frac{3}{4}\right)^3\binom{4}{1}\left(\frac{1}{2}\right)^4 \\ &+ \binom{4}{2}\left(\frac{1}{4}\right)^2\left(\frac{3}{4}\right)^2\binom{4}{2}\left(\frac{1}{2}\right)^4 + \binom{4}{3}\left(\frac{1}{4}\right)^3\left(\frac{3}{4}\right)\binom{4}{3}\left(\frac{1}{2}\right)^4 \\ &+ \binom{4}{4}\left(\frac{1}{4}\right)^4\binom{4}{4}\left(\frac{1}{2}\right)^4 \\ &= \left(\frac{81}{256}\right)\left(\frac{1}{16}\right) + 4\left(\frac{27}{256}\right)4\left(\frac{1}{16}\right) + 6\left(\frac{9}{256}\right)6\left(\frac{1}{16}\right) + 4 \\ &\quad \left(\frac{3}{256}\right)4\left(\frac{1}{16}\right) + \left(\frac{1}{256}\right)\left(\frac{1}{16}\right) \\ &= \frac{443}{2048}. \end{aligned}$$

3. In this case we have

$$\begin{aligned}
P\{X > Y\} &= P\{X=1\}P\{Y=0\} + P\{X=2\}\,[P\{Y=1\}+P\{Y=0\}] \\
&+ P\{X=3\}\,[P\{Y=2\}+P\{Y=1\}+P\{Y=0\}] \\
&+ P\{X=4\}\,[P\{Y=3\}+P\{Y=2\}+P\{Y=1\}+P\{Y=0\}] \\
&= \binom{4}{1}\left(\frac{1}{4}\right)\left(\frac{3}{4}\right)^3 \cdot \binom{4}{0}\left(\frac{1}{2}\right)^4 + \binom{4}{2}\left(\frac{1}{4}\right)^2\left(\frac{3}{4}\right)^2 \cdot \\
&\quad \left[\binom{4}{1}\left(\frac{1}{2}\right)^4 + \binom{4}{0}\left(\frac{1}{2}\right)^4\right] + \binom{4}{3}\left(\frac{1}{4}\right)^3\left(\frac{3}{4}\right) \cdot \\
&\quad \left[\binom{4}{2}\left(\frac{1}{2}\right)^4 + \binom{4}{1}\left(\frac{1}{2}\right)^4 + \binom{4}{0}\left(\frac{1}{2}\right)^4\right] + \binom{4}{4}\left(\frac{1}{4}\right)^4 \cdot \\
&\quad \left[\binom{4}{3}\left(\frac{1}{2}\right)^4 + \binom{4}{2}\left(\frac{1}{2}\right)^4 + \binom{4}{1}\left(\frac{1}{2}\right)^4 + \binom{4}{0}\left(\frac{1}{2}\right)^4\right] \\
&= 4\left(\frac{27}{256}\right)\left(\frac{1}{16}\right) + 6\left(\frac{9}{256}\right)\left[4\left(\frac{1}{16}\right) + \left(\frac{1}{16}\right)\right] + 4\left(\frac{3}{256}\right) \\
&\quad \left[6\left(\frac{1}{16}\right) + 4\left(\frac{1}{16}\right) + \left(\frac{1}{16}\right)\right] + \left(\frac{1}{256}\right) \\
&\quad \left[4\left(\frac{1}{16}\right) + 6\left(\frac{1}{16}\right) + 4\left(\frac{1}{16}\right) + \left(\frac{1}{16}\right)\right] = \frac{525}{4096}.
\end{aligned}$$

4. Here we have

$$\begin{aligned}
P\{X &+ Y \le 5\} = P\{X=0\}[P\{Y=0\}+P\{Y=1\}+P\{Y=2\} \\
&+ P\{Y=3\}+P\{Y=4\}] + P\{X=1\}[P\{Y=0\} \\
&+ P\{Y=1\}+P\{Y=2\}+P\{Y=3\}+P\{Y=4\}] \\
&+ P\{X=2\}\,[P\{Y=0\}+P\{Y=1\}+P\{Y=2\} \\
&+ P\{Y=3\}] + P\{X=3\}\,[P\{Y=0\}+P\{Y=1\} \\
&+ P\{Y=2\}] + P\{X=4\}\,[P\{Y=0\}+P\{Y=1\}]
\end{aligned}$$

$$= \left(\frac{3}{4}\right)^4 \left[\left(\frac{1}{2}\right)^4 + 4\left(\frac{1}{2}\right)^4 + 6\left(\frac{1}{2}\right)^4 + 4\left(\frac{1}{2}\right)^4 + \left(\frac{1}{2}\right)^4\right]$$

$$+ \; 4\left(\frac{1}{4}\right)\left(\frac{3}{4}\right)^3 \left[\left(\frac{1}{2}\right)^4 + 4\left(\frac{1}{2}\right)^4 + 6\left(\frac{1}{2}\right)^4 + 4\left(\frac{1}{2}\right)^4 + \left(\frac{1}{2}\right)^4\right]$$

$$+ \; 6\left(\frac{1}{4}\right)^2\left(\frac{3}{4}\right)^2 \left[\left(\frac{1}{2}\right)^4 + 4\left(\frac{1}{2}\right)^4 + 6\left(\frac{1}{2}\right)^4 + 4\left(\frac{1}{2}\right)^4\right]$$

$$+ \; 4\left(\frac{1}{4}\right)^3\left(\frac{3}{4}\right) \left[\left(\frac{1}{2}\right)^4 + 4\left(\frac{1}{2}\right)^4 + 6\left(\frac{1}{2}\right)^4\right]$$

$$+ \; \left(\frac{1}{4}\right)^4 \left[\left(\frac{1}{2}\right)^2 + 4\left(\frac{1}{2}\right)^4\right] = \frac{3971}{4096}.$$

Example 6.5 A pair of fair die are rolled. Obtain the probabilities of getting 7, 4 or 12.

Solution

	1	2	3	4	5	6
1	2	3	4	5	6	7
2	3	4	5	6	7	8
3	4	5	6	7	8	9
4	5	6	7	8	9	10
5	6	7	8	9	10	11
6	7	8	9	10	11	12

$$P(\chi = 7) = \frac{6}{36} = \frac{1}{6}$$

$$P(\chi = 4) = \frac{3}{36} = \frac{1}{12}$$

$$P(\chi = 12) = \frac{1}{36}.$$

Uniform random variable

This is a continuous random variable taking values between a and b with equal probabilities over intervals of equal length. The probability density function of the uniform random variable is given in Figure 6.4.

$$f_X(x) = \begin{cases} \frac{1}{b-a}, & a < x < b \\ 0, & \text{otherwise.} \end{cases} \tag{6.2.12}$$

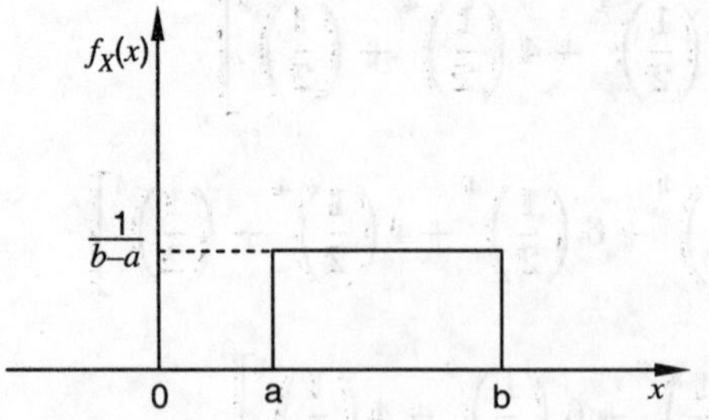

Figure 6.4 *The PDF for uniform random variable*

The uniform random variable is a model for continuous random variables whose range is known, but nothing else is known about the likelihood of various values that the random variable can assume. For example, when the phase of a sinusoid is random, it is usually modelled as a uniform random variable between 0 and 2π.

The MATLAB function *rand*(.) can produce a uniformly distributed random number in the limits $(0, 1)$. For example, if one wants to generate 10 uniformly distributed random numbers as a row vector, the following MATLAB code may be used:

```
y = rand(1,10);
```

The mean and standard deviation of the random numbers can be calculated using the functions *mean*(.) and *std*(.), respectively. The *rand*(.) function can be suitably modified to get any uniformly distributed random numbers.

Gaussian or normal random variable

This is a continuous random variable described by the probability density function

$$f_X(x) = \frac{1}{\sqrt{2\pi}\sigma} e^{-\frac{(x-m)^2}{2\sigma^2}}. \tag{6.2.13}$$

The Gaussian random variable is the most important and frequently used random variable in communications system simulations. The reason is that thermal noise, which is the major source of noise in communication systems, has a Gaussian distribution. The probability density function for the Gaussian random variable with zero mean and variance 0.8 is illustrated in Figure 6.5. The Gaussian PDF is characterised by the 'bell-shaped' curve. The MATLAB script m-file to generate it is also given below.

```
% M-file to plot Gaussian pdf..
x=[-6:.01:6];
sig=.8;
n=1./(sqrt(2*pi)*sig).*exp(-x.^2/(2.*sig.^2));
plot(x,n); xlabel('x');ylabel('Gaussian PDF');
title('The Gaussian PDF for zero mean and 0.8 variance');\\
```

The MATLAB function *randn*(.) generates Gaussian random numbers with zero mean and unit variance.

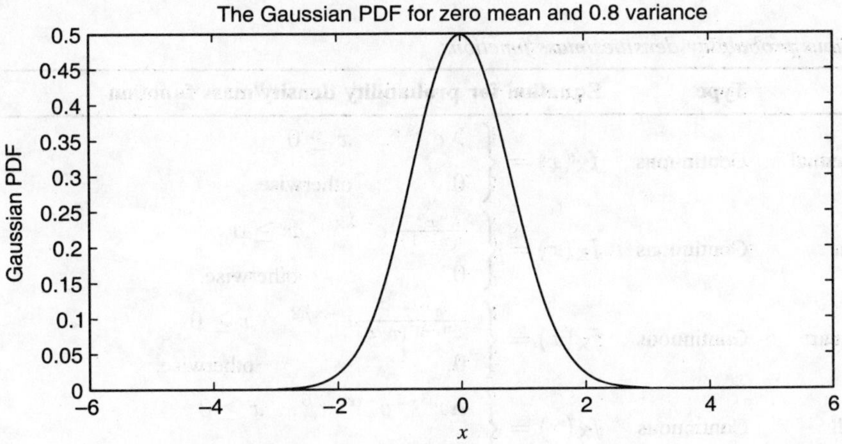

Figure 6.5 *The PDF for Gaussian random variable*

There are several other PDFs of importance, which are summarised in Table 6.1.

The beta function $B(\alpha, \beta)$ is defined as

$$B(\alpha, \beta) = \int_0^1 x^{\alpha-1}(1-x)^{\beta-1}dx = 2\int_0^{2\pi} (\sin\theta)^{2\alpha-1}(\cos\theta)^{2\beta-1}d\theta. \tag{6.2.14}$$

The mean and variance of the probability distributions given in Table 6.1 and other common distributions are tabulated in Table 6.2.

The various probability density functions are illustrated in Figures 6.6 to 6.19. The MATLAB script m-files to generate the plots are also appended.

```
% MATLAB script m-file to generate and
% plot the exponential p.d.f.
clf;
clear all;
lambda=.5;
x=[0:.01:10];
expdf=lambda*exp(-lambda*x);
plot(x,expdf);xlabel('x'); ylabel('p.d.f');
title('The exponential p.d.f. with mean=0.5');
%%%%%%%%%%%%
```

```
%% MATLAB script m-file to generate and
% plot Gamma p.d.f. for various values of
% Lambda & n..
clf;
clear all;
n=2;
```

Table 6.1 *Various probability densities/mass functions*

SNo.	Name	Type	Equation for probability density/mass function
1.	Exponential	Continuous	$f_X(x) = \begin{cases} \lambda e^{-\lambda x}, & x \geq 0 \\ 0, & \text{otherwise.} \end{cases}$
2.	Gamma	Continuous	$f_X(x) = \begin{cases} \frac{\lambda^n x^{n-1}}{(n-1)!} e^{-\lambda x}, & x \geq 0 \\ 0, & \text{otherwise.} \end{cases}$
3.	Chi-square	Continuous	$f_X(x) = \begin{cases} \frac{x^{n/2-1}}{2^{n/2}\Gamma(n/2)} e^{-x/2}, & x \geq 0 \\ 0, & \text{otherwise.} \end{cases}$
4.	Weibull	Continuous	$f_X(x) = \begin{cases} \alpha x^{\beta-1} e^{-\alpha x^\beta/\beta}, & x \geq 0 \\ 0, & \text{otherwise.} \end{cases}$
5.	Rayleigh	Continuous	$f_X(x) = \begin{cases} \frac{x}{\sigma^2} e^{-x^2/2\sigma^2}, & x \geq 0 \\ 0, & \text{otherwise.} \end{cases}$
6.	Nakagami-m	Continuous	$f_X(x) = \begin{cases} \frac{2}{\Gamma(m)} \left(\frac{m}{\Omega}\right)^m x^{2m-1} e^{-mx^2/\Omega}, & x > 0. \\ 0, & \text{otherwise.} \end{cases}$
7.	Beta	Continuous	$f_X(x) = \begin{cases} \frac{1}{B(\alpha,\beta)} x^{\alpha-1}(1-x)^{\beta-1}, & 0 < x < b \\ 0, & \text{otherwise.} \end{cases}$
8.	Cauchy	Continuous	$f_X(x) = \frac{\alpha/\pi}{(x-\mu)^2+\alpha^2}, \quad \lvert x \rvert < \infty$
9.	Laplace	Continuous	$f_X(x) = \frac{\alpha}{2} e^{-\alpha \lvert x \rvert}, \quad \lvert x \rvert < \infty$
10.	Maxwell	Continuous	$f_X(x) = \begin{cases} \frac{4}{\alpha^3\sqrt{\pi}} x^2 e^{-x^2/\alpha^2}, & x \geq 0 \\ 0, & otherwise. \end{cases}$
11.	Poisson	Discrete	$P\{X = k\} = e^{-\lambda} \frac{\lambda^k}{k!}, \quad k = 0, 1, 2, \ldots, \infty$
12.	Geometric	Discrete	$P\{X = k\} = pq^{k-1}, \quad k = 1, 2, 3, \ldots, \infty$
13.	Neg. binomial	Discrete	$P\{X = k\} = \binom{k-1}{r-11} p^r q^{k-r},$ $k = r, r+1, \ldots, \infty$
14.	Discrete uniform	Discrete	$P\{X = k\} = \frac{1}{N}, \quad k = 1, 2, \ldots, N.$

Table 6.2 *Means/variances of probability densities/mass functions*

Sl.No.	Name	Mean	Variance
1.	Normal	μ	σ^2
2.	Exponential	$\frac{1}{\lambda}$	$\frac{1}{\lambda^2}$
3.	Gamma $G(\alpha, \beta)$	$\alpha\beta$	$\alpha\beta^2$
4.	Chi-square $\chi^2(n)$	n	2n
5.	Weibull	$\left(\frac{\beta}{\alpha}\right)^{1/\beta}\Gamma\left(1+\frac{1}{\beta}\right)$	$\left(\frac{\beta}{\alpha}\right)^{2/\beta}\left[\Gamma\left(1+\frac{2}{\beta}\right)-\left(\Gamma\left(1+\frac{1}{\beta}\right)\right)^2\right]$
6.	Rayleigh	$\sqrt{\frac{\pi}{2}}\sigma$	$(2-\pi/2)\sigma^2$
7.	Nakagami-m	$\frac{\Gamma(m+1/2)}{\Gamma(m)}\sqrt{\frac{\Omega}{m}}$	$\Omega\left(1-\frac{1}{m}\left(\frac{\Gamma(m+1/2)}{\Gamma(m)}\right)^2\right)$
8.	Uniform $U(a, b)$	$\frac{a+b}{2}$	$\frac{(b-a)^2}{12}$
9.	Beta $\beta(\alpha, \beta)$	$\frac{\alpha}{\alpha+\beta}$	$\frac{\alpha\beta}{(\alpha+\beta)^2(\alpha+\beta+1)}$
10.	Cauchy	—	∞
11.	Laplace	$\frac{1}{\alpha}$	$\frac{2-\alpha}{\alpha^3}$
12.	Maxwell	$2\alpha\sqrt{2/\pi}$	$(3-\frac{8}{\pi})\alpha^2$
13.	Poisson	λ	λ
14.	Geometric	$\frac{1}{p}$	$\frac{1-p}{p^2}$
15.	Bernoulli	p	$p(1-p)$
16.	Binomial	np	$np(1-p)$
17.	Neg. binomial	$\frac{r}{p}$	$\frac{r(1-p)}{p^2}$
18.	Discrete uniform	$\frac{N+1}{2}$	$\frac{N^2-1}{12}$

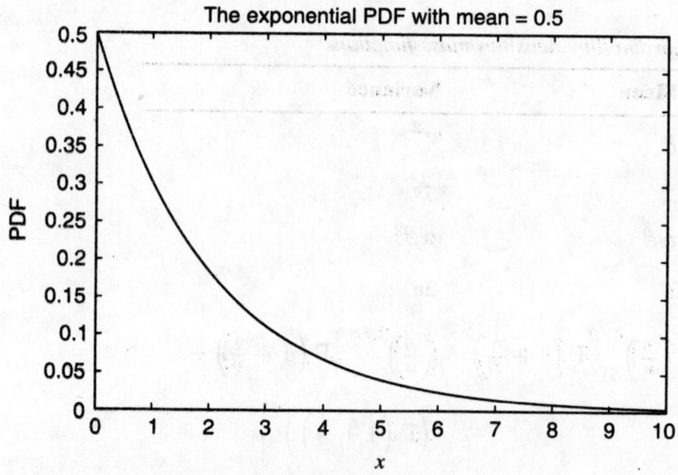

Figure 6.6 *The PDF for exponential random variable*

```
beta=[0.5,1,2,4];
lambda=1./beta;
x=[0:.01:10];
for i=1:4
  gampdf=(lambda(i)^n*x.^(n-1)/prod(1:(n-1)))...
    .*exp(-lambda(i)*x);
  plot(x,gampdf); hold on;
end;
xlabel('x'); ylabel('Gamma p.d.f.');
title('The Gamma pdf with n=2');
hold off;
%%%%%%%%

%% MATLAB script m-file to generate and
% plot Chi-Square p.d.f. for various values of
% n..
clf;
clear all;
n=[2,6,8,10];
x=[0:.01:30];
for i=1:4
  chpdf= x.^(n(i)/2-1).*exp(-x/2)/(2^(n(i)/2)...
    *prod(1:(n(i)/2 -1)));
  plot(x,chpdf); hold on;
end;
xlabel('x'); ylabel('f_X(x)');
title('The \chi^2 pdf with n=2, 6, 8 & 10');
hold off;
%%%%%
```

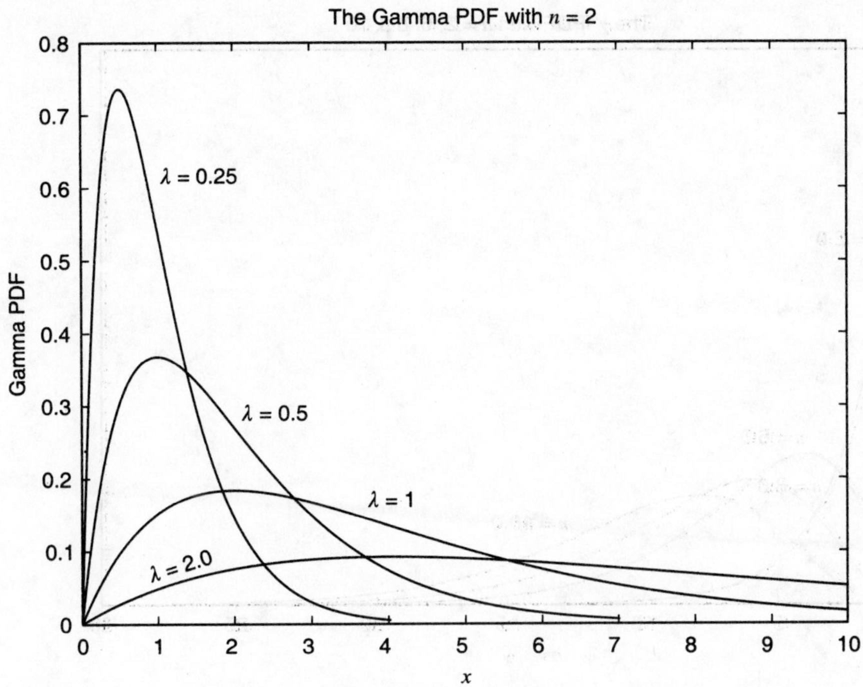

Figure 6.7 *The PDF for Gamma random variable*

```
%% MATLAB script m-file to generate and
% plot Weibull p.d.f. for various values of
% alpha & beta..
clf;
clear all;
a=[0.5,2,4];
b=[2,2,4];
x=[0:.01:10];
for i=1:3
  wepdf=a(i)*x.^(b(i)-1).*exp(-a(i).*x.^b(i)/b(i));
  plot(x,wepdf); hold on;
end;
xlabel('x'); ylabel('f_X(x)');
title('The Weibull pdf');
hold off;
%%%%
```

```
%% MATLAB script m-file to generate and
% plot Rayleigh p.d.f. for various values of
% sigma..
clf;
clear all;
```

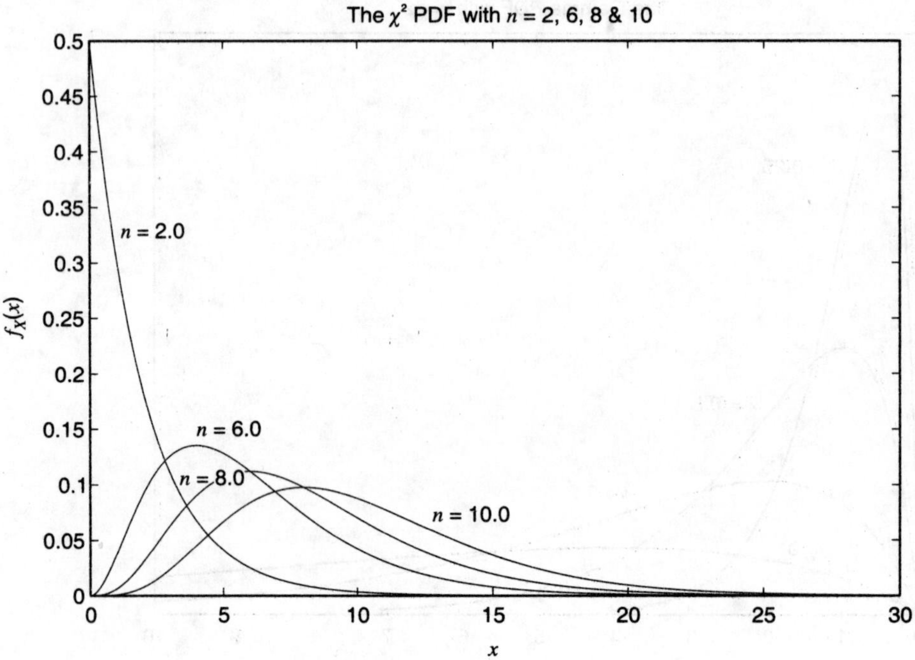

Figure 6.8 *The PDF for χ^2 random variable*

```
a=[16,9,4,1];
b=2;
x=[0:.01:4];
for i=1:4
  rpdf=a(i)*x.^(b-1).*exp(-a(i).*x.^b/b);
  plot(x,rpdf); hold on;
end;
xlabel('x'); ylabel('f_X(x)');
title('The Rayleigh pdf');
hold off;
%%%%%%%%%%%

%% MATLAB script m-file to generate and
% plot Nakagami-m p.d.f. for various values of
% m..
clf;
clear all;
m=[.25,.5,.75,1,2,4];
O=.08;
x=[.01:.01:2];
for i=1:6
  nadf=2*m(i)^m(i).* x.^(2*m(i)-1).*exp(-m(i)...
    *x.^2/O)/(prod(1:(m(i)-1))*O^m(i));
```

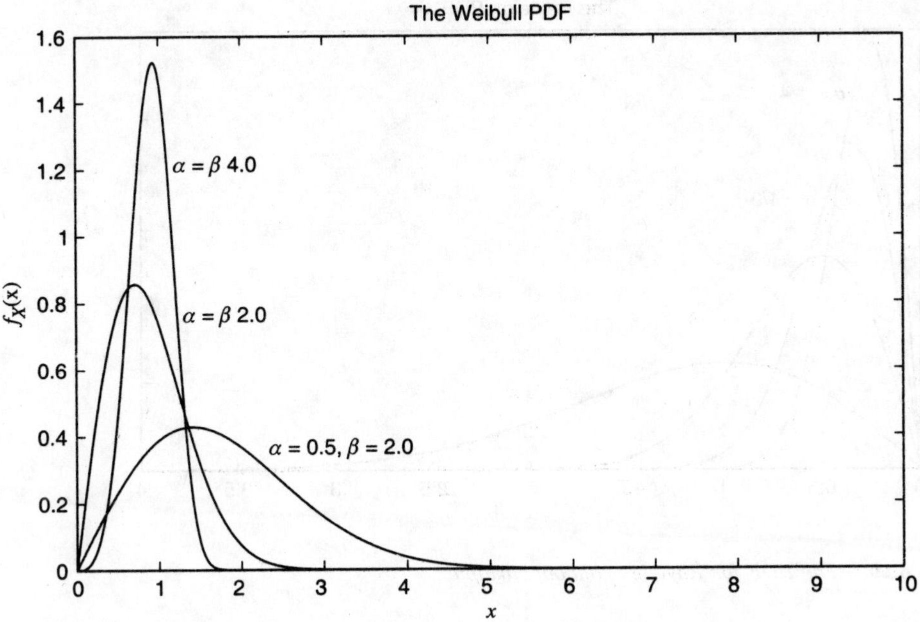

Figure 6.9 *The PDF for Weibull random variable*

```
  plot(x,nadf); hold on; axis([.01 2 0 10]);
end;
xlabel('x'); ylabel('f_X(x)');
title('The Nakagami-m pdf');
hold off;
%%%%%%%%%%

%% MATLAB script m-file to generate and
% plot the Beta p.d.f. for various values of
% alpha and beta.
clf;
clear all;
a=[.5 3 8];
b=[.5 3 2];
x=[.01:.01:1];
for i=1:3
  bedf=x.^(a(i)-1).*(1-x).^(b(i)-1)/beta(a(i),b(i));
  plot(x,bedf); hold on; axis([.01 1 0 5]);
end;
xlabel('x'); ylabel('f_X(x)');
title('The Beta pdf');
hold off;
%%%%%%%%%%%%
```

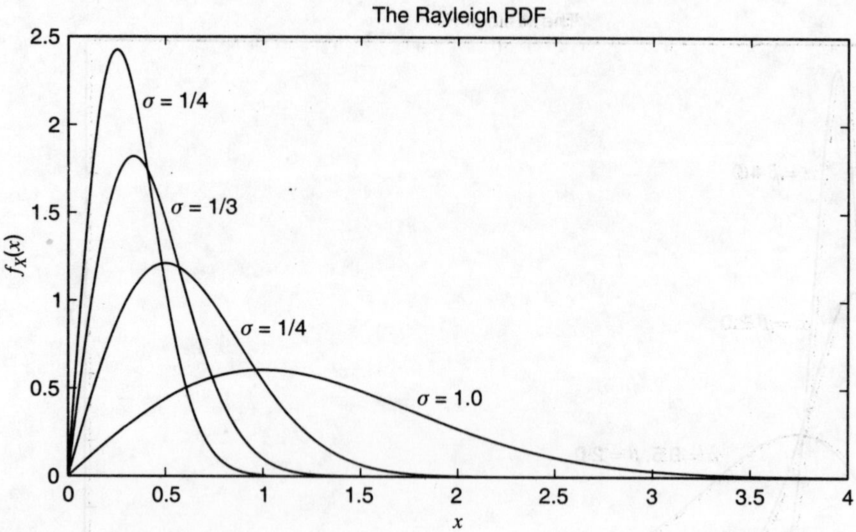

Figure 6.10 *The PDF for Rayleigh random variable*

```
%% MATLAB script m-file to generate and
% plot the Cauchy p.d.f. for various values of
% alpha and mu.
clf;
clear all;
a=[10*pi 40*pi];
m=[0 8];
x=[-1000:1000];
for i=1:2
  cadf= a(i)/pi./((x-m(i)).^2+a(i)^2);
  plot(x,cadf); hold on;
end;
xlabel('x'); ylabel('f_X(x)');
title('The Cauchy pdf');
hold off;
%%%%%%%%%%

%% MATLAB script m-file to generate and
% plot the Laplace p.d.f. for various values of
% alpha ...
clf;
clear all;
a=[.4:.8:1.2];
x=[-20:.01:20];
for i=1:2
  ladf= a(i).*exp(-a(i).*abs(x))/2;
  plot(x,ladf); hold on; axis([-20 20 0 .8]);
```

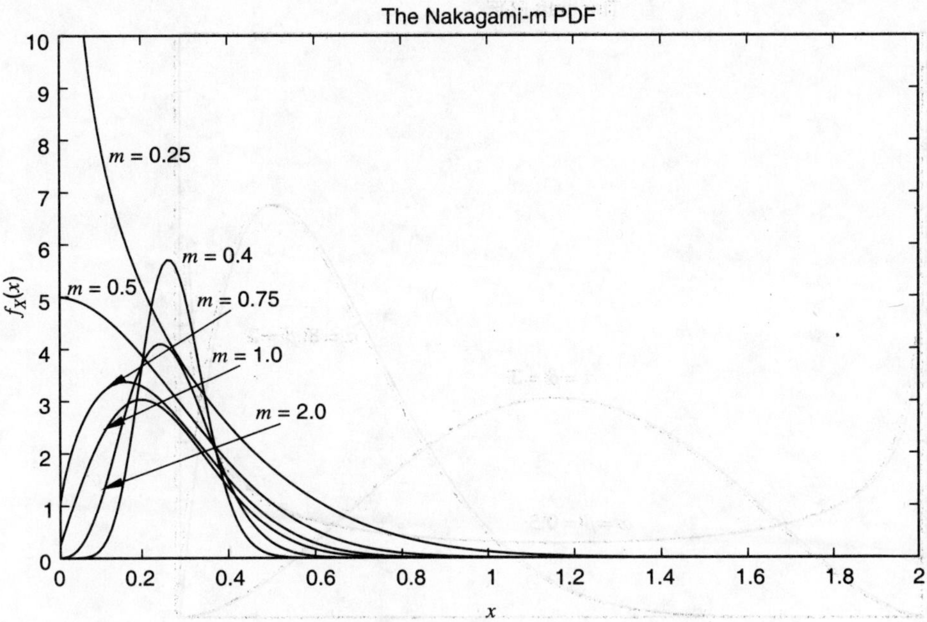

Figure 6.11 *The PDF for Nakagami-m random variable*

```
end;
xlabel('x'); ylabel('f_X(x)');
title('The Laplace pdf');
hold off;
%%%%%%%%%%

%% MATLAB script m-file to generate and
% plot the Maxwell p.d.f. for various values of
% alpha ...
clf;
clear all;
a=[1:4];
x=[0:.01:10];
for i=1:4
  madf= 4*x.^2.*exp(-x.^2/a(i)^2)/(a(i)^3*sqrt(pi));
  plot(x,madf); hold on;
end;
xlabel('x'); ylabel('f_X(x)');
title('The Maxwell pdf');
hold off;
%%%%%%%%%%%

%% MATLAB script m-file to generate and
% plot the Poisson  p.m.f. for various values of
```

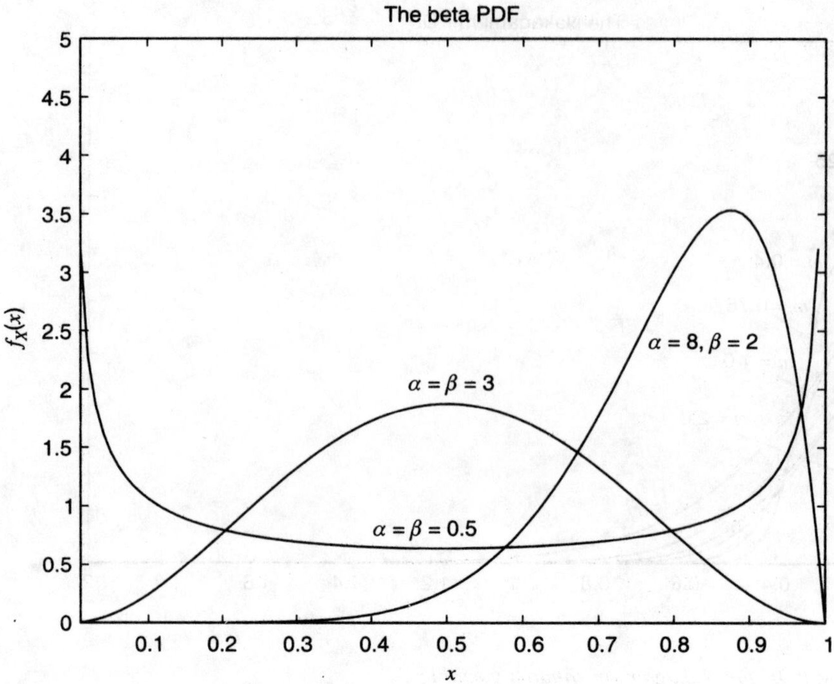

Figure 6.12 *The PDF for beta random variable*

```
% alpha ...
clf;
clear all;
l=3;
k=[0:10];
ppmf=exp(-l)*l.^k./factorial(k);
stem(k,ppmf);
xlabel('k'); ylabel('P\{X = k\}');
title('The Poisson pmf');
hold off;
%%%%%%%%%%%%%%

%% MATLAB script m-file to generate and
% plot the Negative Binomial p.m.f. for various values of
% p...
clf;
clear all;
p= .2;
r=2;
for k=r:30
npmf= nchoosek(k-1,r-1)*p^r*(1-p)^(k-r);
stem(k,npmf); hold on;
```

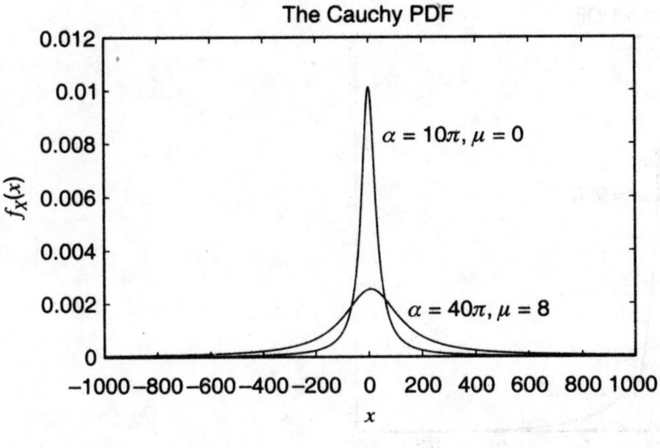

Figure 6.13 *The PDF for Cauchy random variable*

```
end;
xlabel('k'); ylabel('P\{X = k\}');
title('The Negative Binomial pmf for p=0.5 and r=2');
hold off;
%%%%%%%%%%%%%%%%%%%%%%%%%%%%%%%%%

%% MATLAB script m-file to generate and
% plot the Geometric  p.m.f. for various values of
% p...
clf;
clear all;
p= .5;
k=[1:10];
gpmf= p*(1-p).^(k-1);
stem(k,gpmf);
xlabel('k'); ylabel('P\{X = k\}');
title('The Geometric pmf for p=0.5');
hold off;
%%%%%%%%%%%%%%%%%%%%%%%%%%%%%%%%%

%% MATLAB script m-file to generate and
% plot the Discrete Normal p.m.f.
clf;
clear all;
k=1:10;
dn= .1*[ones(size(k))];
stem(k,dn); axis([-1 12 0 .2]);
xlabel('k'); ylabel('P\{X = k\}');
title('The Discrete Normal pmf for k=1 to 10');
```

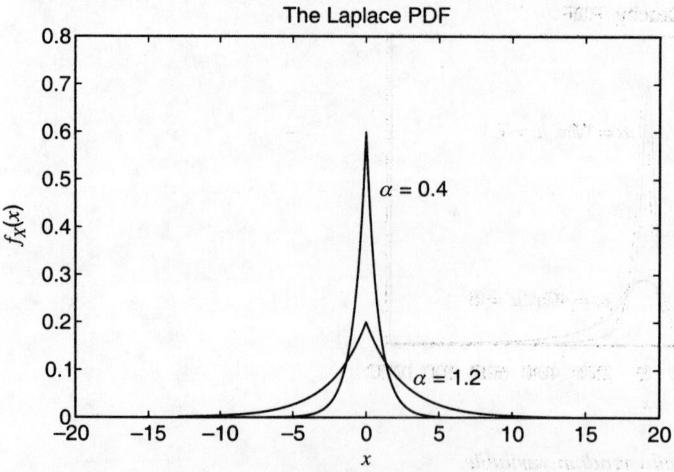

Figure 6.14 *The PDF for Laplace random variable*

The $Q(x)$ function

The cumulative distribution function for the Gaussian random variable with mean $m = 0$ and variance $\sigma^2 = 1$is denoted by $\Phi(x)$ and is given by

$$\Phi(x) = P\{X \leq x\} = \int_{-\infty}^{x} \frac{1}{\sqrt{2\pi}} e^{-\frac{t^2}{2}} dt. \tag{6.2.15}$$

A closely related function is $Q(x) = 1 - \Phi(x)$ giving $P\{X > x\}$. This function is well tabulated and quite often used in communications system simulations. We can show that $Q(x)$ satisfies the following relations:

$$Q(-x) = 1 - Q(x) \tag{6.2.16}$$

$$Q(0) = \frac{1}{2} \tag{6.2.17}$$

$$Q(\infty) = 0. \tag{6.2.18}$$

The reader is invited to attempt prove these results.

The erf(.) function

The Gaussian CDF defined in terms of the error function (erf) and complementary error function (erfc). The *erf*(.) is defined as

$$\text{erf}(u) = \frac{2}{\sqrt{\pi}} \int_{0}^{u} e^{-z^2} dz = -\text{erf}(-u). \tag{6.2.19}$$

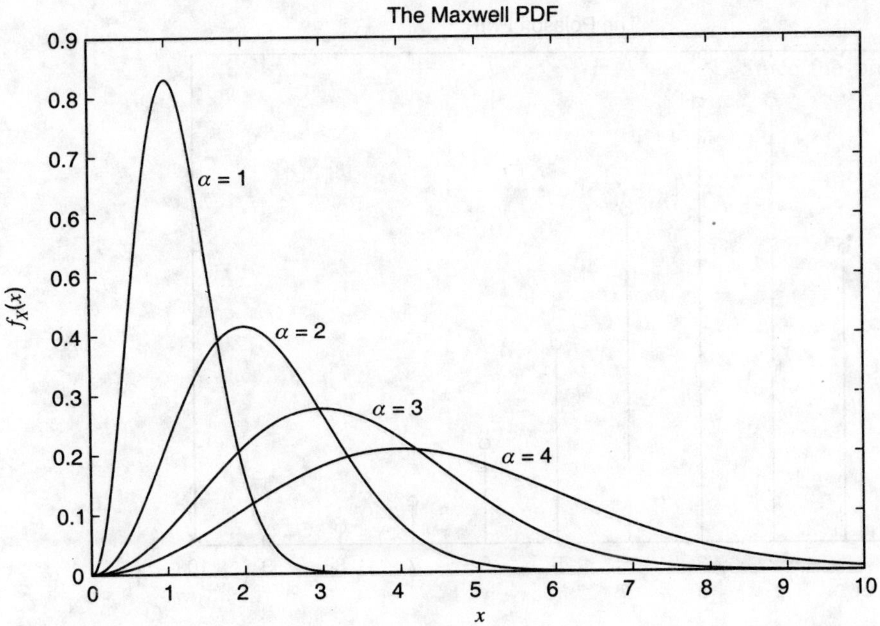

Figure 6.15 *The PDF for Maxwell random variable*

The complementary error functionis defined as erfc(u) = 1−erf(u). Now for a Gaussian random variable X with mean μ_X and standard deviation σ_X:

$$P(X \leq x) = \frac{1}{2}\left[1 + \text{erf}\left(\frac{x - \mu_X}{\sqrt{2}\sigma_X}\right)\right]. \tag{6.2.20}$$

For example, for $X \sim \mathcal{N}(0,1)$ (Gaussian random variable with 0 mean and unity variance),

$$\begin{aligned} P(X > x) &= 1 - P(X \leq x) = 1 - \frac{1}{2}\left[1 + \text{erf}\left(\frac{x}{\sqrt{2}}\right)\right] \\ &= 0.5 - 0.5\,\text{erf}\left(\frac{x}{\sqrt{2}}\right) = 0.5\,\text{erfc}\left(\frac{x}{\sqrt{2}}\right). \end{aligned}$$

For $\quad X \sim \mathcal{N}(0,1)$,

$$Q(x) \triangleq P(X > x) = 0.5\,\text{erfc}\left(\frac{x}{\sqrt{2}}\right).$$

Then for $\quad Z \sim \mathcal{N}\left(\mu_Z, \sigma_Z^2\right)$,

$$P(Z > z) = Q\left(\frac{z - \mu_Z}{\sigma_Z}\right).$$

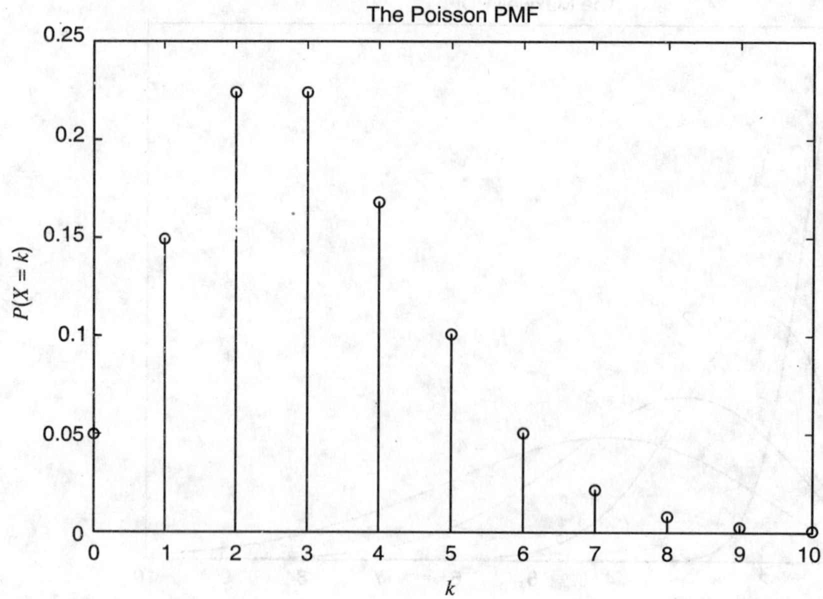

Figure 6.16 *The PMF for Poisson random variable*

6.2.4 Functions of a random variable

A function of a random variable $Y = g(X)$ is itself a random variable. To find its cumulative distribution function we can use the definition of the CDF to get

$$F_Y(y) = P\{\omega \in \Omega;\ \ g(X(\omega)) \le x\}. \tag{6.2.21}$$

In the special case that, for all y, the equation $g(x) = y$ has a countable number of solutions $\{x_i\}$ and for all these solutions $g^{'}(x_i)$ exists and is nonzero, we can use the following relation to find the probability density function of $Y = g(X)$:

$$f_Y(y) = \sum_i \frac{f_X(x_i)}{|g^{'}(x_i)|}. \tag{6.2.22}$$

6.2.5 Jointly distributed random variables

Let $X(x)$ and $Y(y)$ be two random variables. Then their *joint probability distribution* is given by

$$f_{XY} = \frac{\delta^2 F_{XY}(x,y)}{\delta x\, \delta y}. \tag{6.2.23}$$

The respective single distributions can be recovered as

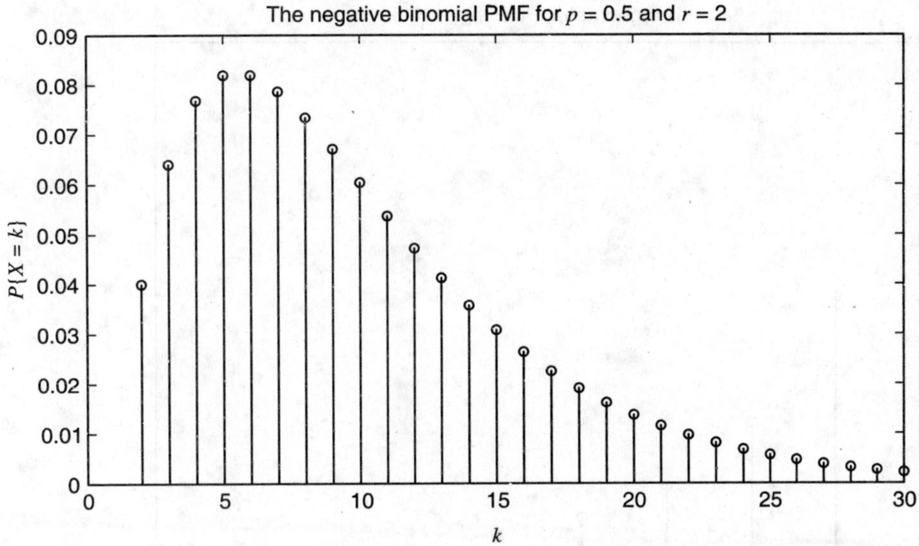

Figure 6.17 *The PMF for negative binomial random variable*

$$f_X(x) = \int_{-\infty}^{\infty} f_{XY}(x,y)dy, \tag{6.2.24}$$

$$f_Y(y) = \int_{-\infty}^{\infty} f_{XY}(x,y)dx. \tag{6.2.25}$$

Conditional probability density function

The conditional probability density function of random variable X, given the outcome of random variable Y is given by

$$f_{X|Y}(x|Y=y) = \frac{f_{XY}(x,y)}{f_Y(y)}. \tag{6.2.26}$$

Statistical independence

The random variables X and Y are statistically independent if and only if

$$f_{X|Y}(x|Y=y) = f_X(x), \tag{6.2.27}$$

$$f_{XY}(x,y) = f_X(x)\, f_Y(x). \tag{6.2.28}$$

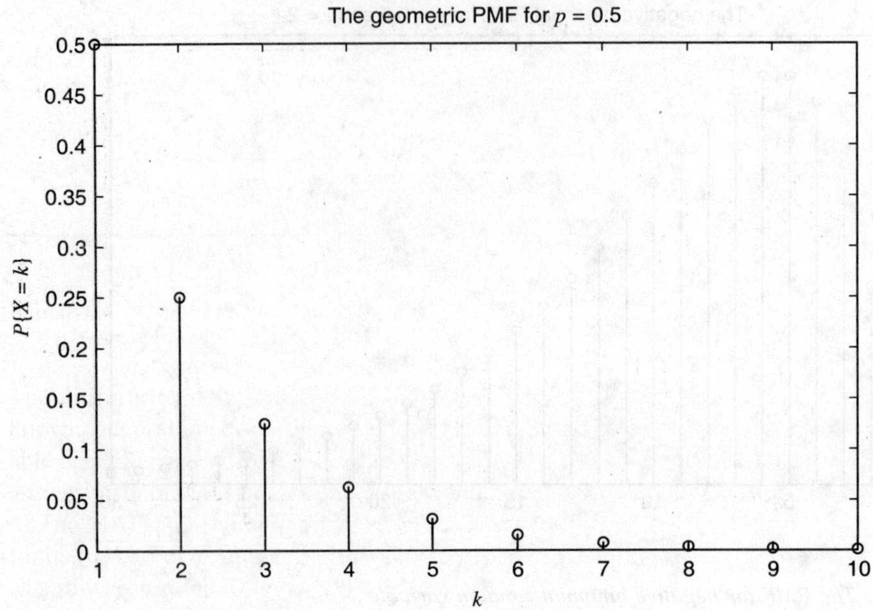

Figure 6.18 *The PMF for geometric random variable*

6.2.6 Functions of several random variables

The results derived in the preceding section can be generalised to fine the joint distribution of n random variables $Y_1, Y_2, \ldots, Y_n$, given the distributions of n related random variables $X_1, X_2, \ldots, X_n$ and the relationship between the two sets of random variables.

$$Y_i = g_i(X_1, X_2, \ldots, X_n), \quad i = 1, 2, \ldots, n.$$

Using the vector notation, it can be shown that

$$f_Y(y) = \sum_{i=1}^{k} \frac{f_X(x^{(i)})}{|J(x^{(i)})|}, \tag{6.2.29}$$

where $x^{(i)} = [x_1^{(i)}, x_2^{(i)}, \ldots, x_n^{(i)}]^T$ is the ith solution to $y = g(x) = [g_1(x), g_2(x), \ldots, g_n(x)]^T$, and the Jacobian J is defined as

$$J[x^{(i)}] = \begin{vmatrix} \frac{\partial g_1}{\partial x_1} & \frac{\partial g_1}{\partial x_2} & \cdots & \frac{\partial g_1}{\partial x_m} \\ \vdots & \vdots & & \vdots \\ \frac{\partial g_n}{\partial x_1} & \frac{\partial g_n}{\partial x_2} & \cdots & \frac{\partial g_n}{\partial x_m} \end{vmatrix}_{at\ x^{(i)}}. \tag{6.2.30}$$

Suppose we have n random variables with known PDF, and we are interested in the joint PDF of $m < n$ functions of them, say

$$y_i = g_i(x_1, x_2, \ldots, x_n), \quad i = 1, 2, \ldots, m.$$

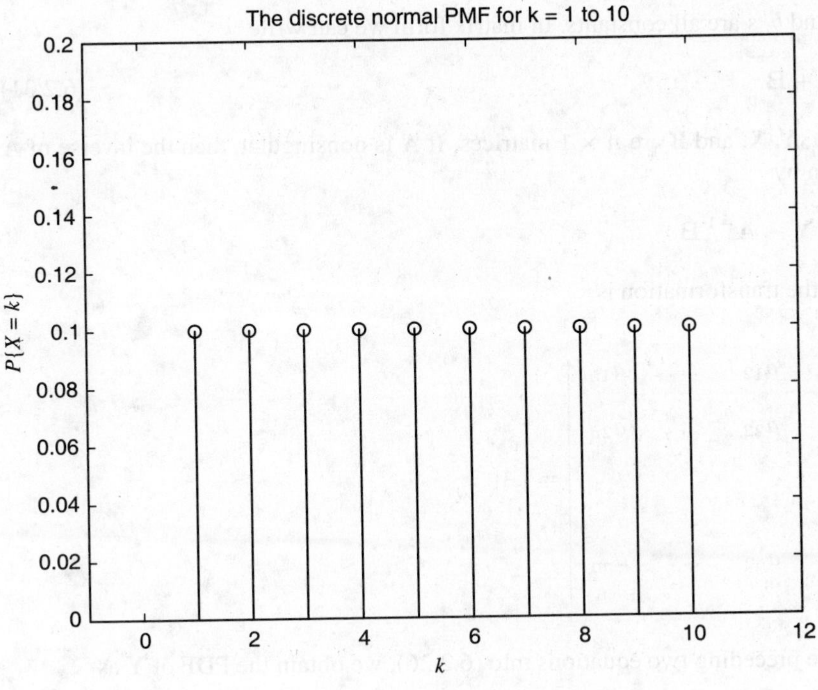

Figure 6.19 *The PMF for discrete normal random variable*

Now, we can define $n - m$ additional functions

$$y_j = g_j(x_1, x_2, \ldots, x_n), \quad j = m + 1, \ldots, n$$

in any convenient way so that the Jacobian is nonzero, compute the joint PDF of $Y_1, Y_2, \ldots, Y_n$, and then obtain the marginal PDF of $Y_1, Y_2, \ldots, Y_m$ by integrating $Y_{m+1}, \ldots, Y_n$. If the additional functions are carefully chosen, then the inverse can easily be found and the resulting integration can be handled, but often with some difficulty.

Special case linear transformation

One of the most frequently used types of transformations is the *affine transformation*, where each of the new variables is a linear combination of the old variables plus a constant. That is,

$$\begin{aligned}
Y_1 &= a_{11}X_1 + a_{12}X_2 + \cdots + a_{1n}X_n + b_1 \\
Y_2 &= a_{21}X_1 + a_{22}X_2 + \cdots + a_{2n}X_n + b_2 \\
&\vdots \\
Y_n &= a_{n1}X_1 + a_{n2}X_2 + \cdots + a_{nn}X_n + b_n
\end{aligned}$$

where the a_{ij} 's and b_i s are all constants. In matrix form we can write

$$\mathbf{Y} = \mathbf{A}\,\mathbf{X} + \mathbf{B}, \tag{6.2.31}$$

where **A** is $n \times n$, **Y, X,** and **B** are $n \times 1$ matrices. If **A** is nonsingular, then the inverse of **A** exists and is given by

$$\mathbf{X} = \mathbf{A}^{-1}\mathbf{Y} - \mathbf{A}^{-1}\mathbf{B}.$$

The Jacobian of the transformation is

$$J = \begin{vmatrix} a_{11} & a_{12} & \cdots & a_{1n} \\ a_{21} & a_{22} & \cdots & a_{2n} \\ \vdots & \vdots & & \vdots \\ a_{n1} & a_{n2} & \cdots & a_{nn} \end{vmatrix} = |A|.$$

Substituting the preceding two equations into (6.2.26), we obtain the PDF of **Y** as

$$\mathbf{f_Y(y) = f_X(A^{-1}y - A^{-1}B)\|A\|^{-1}}. \tag{6.2.32}$$

Sum of random variables

Let us now consider $Y_1 = X_1 + X_2$, where X_1 and X_2 are independent random variables. We introduce an additional random variable $Y_2 = X_2$ so that the transformation is given by

$$\begin{bmatrix} Y_1 \\ Y_2 \end{bmatrix} = \begin{bmatrix} 1 & 1 \\ 0 & 1 \end{bmatrix} \begin{bmatrix} X_1 \\ X_2 \end{bmatrix}.$$

From (6.2.26) it follows that

$$\begin{aligned} f_{Y_1,Y_2}(y_1, y_2) &= f_{X_1,X_2}(y_1 - y_2, y_2) \\ &= f_{X_1}(y_1 - y_2).f_{X_2}(y_2) \end{aligned}$$

since X_1 and X_2 are independent. The PDF of Y_1 is obtained by integration.

$$f_{Y_1}(y_1) = \int_{-\infty}^{\infty} f_{X_1}(y_1 - y_2).f_{X_2}(y_2)dy_2. \tag{6.2.33}$$

The relationship given in (6.2.33) is the convolution of f_{X_1} and f_{X_2}, which is written as

$$f_{Y_1} = f_{X_1} * f_{X_2}. \tag{6.2.34}$$

Thus, we have:

The density function of the sum of two independent random variables is given by the convolution of their densities.

It can be shown that if X_1 and X_2 are Gaussian, then $Y_1 = X_1 + X_2$ will also be Gaussian. This result cannot be generalised to non-Gaussian densities.

Example 6.6 Assuming X is a Gaussian random variable with $m = 0$ and $\sigma = 1$, find the PDF of the random variable Y given by $Y = aX + b$.

Solution
In this case $g(x) = ax + b$, and hence $g'(x) = a$. The equation $ax + b = y$ has only one solution given by $x_1 = \frac{y-b}{a}$. Using these results, we obtain

$$
\begin{aligned}
f_Y(y) &= \frac{f_X(\frac{y-b}{a})}{|a|} \\
&= \frac{1}{\sqrt{2\pi a^2}} e^{-\frac{(y-b)^2}{2a^2}}.
\end{aligned}
$$

It may be noted that Y is a Gaussian random variable with mean b and variance a^2. It can be denoted as $\mathcal{N}(b, a^2)$. From the preceding example we arrive at the important conclusion that:

A linear function of a Gaussian random variable is itself a Gaussian random variable.

6.2.7 Statistical averages

Several averages are defined on random variables.

Expected value or mean

The *expected value or mean* of the random variable X is defined as

$$E(X) \triangleq \int_{-\infty}^{\infty} x f_X(x) dx. \tag{6.2.35}$$

The expected value of $Y = g(X)$ is given by

$$E(Y) = E(g(X)) = \int_{-\infty}^{\infty} g(x) f_X(x) dx. \tag{6.2.36}$$

For discrete random variables the corresponding equation is

$$E(X) \triangleq \sum_i x_i P\{X = x_i\} \tag{6.2.37}$$

and for a function of the random variable, $Y = g(X)$,

$$E(Y) = E(g(X)) = \sum_{i} g(x_i) P\{X = x_i\}. \tag{6.2.38}$$

Variance and standard deviation

The variance of a random variable X is defined as

$$\begin{aligned}\sigma_X^2 &= E(\{X - E(X)\}^2) = E(X^2 - 2XE(X) + (E(X))^2) \\ &= E(X^2) - 2E(X).E(X) + (E(X))^2 = E(X^2) - \{E(X)\}^2\end{aligned}$$

This relation for variance is obtained since for any constant c, the following hold:

1. $E(cX) = cE(X)$.
2. $E(c) = c$.
3. $E(X + c) = E(X) + c$.

It is also possible to verify the following properties of variance:

1. $\text{VAR}(cX) = c^2\,\text{VAR}(X)$.
2. $\text{VAR}(c) = 0$.
3. $\text{VAR}(X + c) = \text{VAR}(X)$.

The proofs of these relationships are left as an exercise to the reader. Now, by definition, the *standard deviation* is the square root of variance, i.e., σ_X.

Example 6.7 Show that for a Rayleigh density function,

$$f_X(x) = \begin{cases} \dfrac{x}{\sigma^2} e^{-\frac{x^2}{2\sigma^2}}, & x > 0 \\ 0, & \text{otherwise.} \end{cases}$$

$$E[X] = \sigma\sqrt{\frac{\pi}{2}} \text{ and VAR } (X) = (2 - \frac{\pi}{2})\sigma^2.$$

Solution

We have

$$E[X] \triangleq \int_{-\infty}^{\infty} x\, f_X(x)\, dx = \int_{0}^{\infty} \frac{x^2}{\sigma^2} \exp\left(-x^2/2\sigma^2\right) dx.$$

But we have the standard definite integral:

$$\int_{0}^{\infty} x^2 \exp\left(-ax^2\right) dx = \frac{1}{4a}\sqrt{\frac{\pi}{a}}, \quad a > 0 \tag{6.2.39}$$

Hence by putting $a = \frac{1}{2\sigma^2}$,

$$\begin{aligned} E[X] &= \frac{1}{\sigma^2}\int_0^\infty x^2 \exp\left(-ax^2\right) dx \\ &= \frac{1}{\sigma^2} \times \frac{\sigma^2}{2} \times \sqrt{2\pi\sigma^2} = \sigma\sqrt{\frac{\pi}{2}}. \end{aligned}$$

Now the second moment of X is given by

$$E[X^2] \triangleq \int_{-\infty}^{\infty} x^2 f_X(x)\, dx = \int_0^\infty \frac{x^3}{\sigma^2} \exp\left(-x^2/2\sigma^2\right) dx.$$

Now substituting $y = (\frac{x}{\sigma})^2$, we get

$$E[X^2] = \frac{\sigma^2}{2}\int_0^\infty y \exp\left(-y/2\right) dy = 2\sigma^2.$$

Therefore, we have

$$\begin{aligned} \text{VAR}\{X\} &\triangleq E[X^2] - E^2[X] \\ &= 2\sigma^2 - \frac{\pi}{2}\sigma^2 = (2 - \frac{\pi}{2})\sigma^2. \end{aligned}$$

Moments of a random variable

In general, the nth moment of a random variable X is defined as

$$m_X^{(n)} \triangleq \int_{-\infty}^{\infty} x^n f_X(x) dx. \tag{6.2.40}$$

Expected values

Expected values are a calculation of averages. The nth statistical moment is

$$E\left[x^n\right] = \int_{-\infty}^{\infty} x^n f_X(x) dx. \tag{6.2.41}$$

The first moment is the *mean*, the second moment is the *mean square*, the third moment is the *skewness*, and the fourth one is the *kurtosis*.

Characteristic functions

The characteristic function of a random variable X is denoted by $\Psi_X(\omega)$ and is defined as

$$\Psi_X(\omega) \triangleq \int_{-\infty}^{\infty} f_X(x)e^{j\omega x}dx. \tag{6.2.42}$$

The characteristic function of a random variable is closely related to the Fourier transform of its density function. The characteristic function provides a simple means to find the various moments. To obtain the nth moment of the random variable X, we can use the relation

$$m_X^{(n)} = \frac{1}{j^n}\frac{d^n}{d\omega^n}\Psi_X(\omega)\Big|_{\omega=0}. \tag{6.2.43}$$

Example 6.8 The random variable X is $\mathcal{N}(5, 2)$ and $Y = 2X+4$. Find μ_Y, σ_Y and $f_Y(y)$.

Solution

We have, by definition,

$$\begin{aligned}\mu_Y &= E[2X+4]\\ &= 2E[X]+4 = 2\times 5+4 = 14.\end{aligned}$$

We know that $\mu_X = 5$ and $\sigma_X = 2$. Therefore, $E[X^2] = \sigma_X^2 + \mu_X^2 = 4 + 25 = 29$. Now, the second moment of Y is given by

$$\begin{aligned}E[Y^2] &= E[(2X+4)^2] = E[4X^2+16X+16]\\ &= 4E[X^2]+16E[X]+16 = 4\times 29+16\times 5+16 = 212.\end{aligned}$$

Now, $\sigma_Y^2 = E[Y^2] - \mu_Y^2 = 212 - 196 = 16$. Since X is $\mathcal{N}(5, 2)$, $Y$is $\mathcal{N}(14, 16)$. So we can write the PDF of Y as

$$f_Y(y) = \frac{1}{4\sqrt{2\pi}}\exp\left(-\frac{y^2-196}{512}\right).$$

Example 6.9 If X is $\mathcal{N}(0, 4)$ and $Y = 3X^2$, find μ_Y σ_Y, and $f_Y(y)$.

Solution

By definition, $\mu_Y = E[Y] = E[3X^2] = 3E[X^2]$.

But, $\sigma_X^2 = 4 = E[(X-\mu_X)^2] = E[X^2] - \mu_X^2 = E[X^2] = 4$, as $\mu_X = 0$.

Hence, $\mu_Y = 3.4 = 12$.

Now, $\sigma_Y^2 = E[(Y-\mu_Y)^2] = E[Y^2] - \mu_Y^2 = E[9X^4] - 144 = 9E[X^4] - 144$.

The characteristic function of the Gaussian random variable is given by

$$\psi_X(\omega) = \int_{-\infty}^{\infty} e^{j\omega x} f_X(x)dx$$

$$= \frac{1}{2\sqrt{2\pi}} \int_{-\infty}^{\infty} e^{j\omega x} e^{-x^2/(2.4)} dx = e^{-2\omega^2}.$$

Now, $E[X^4] = \frac{1}{j^4}\frac{d^4}{d\omega^4}\psi_X(\omega)|_{\omega=0} = 48e^{-2\omega^2} + \ldots + \ldots|_{\omega=0} = 48.$

Therefore, $\sigma_Y^2 = E[9X^4] - 144 = 9E[X^4] - 144 = 9 \times 48 - 144 = 288.$

The equation $y = 3x^2$ has two solutions, $x_1 = \sqrt{\frac{y}{3}}$ and $x_2 = -\sqrt{\frac{y}{3}}$. Substituting the two values in (6.2.22),

$$f_Y(y) = \frac{1}{6\sqrt{y/3}}\left[f_X\left(\sqrt{\frac{y}{3}}\right) + f_X\left(-\sqrt{\frac{y}{3}}\right)\right], \quad y > 0.$$

$$= \frac{e^{-y/(2.3.4)}}{2\sqrt{2\pi 3y}}, \quad y > 0.$$

Note that $f_Y(y)$ is a χ^2 distribution with $n = 1$ so that $\Gamma(n/2) = \Gamma(1/2) = \sqrt{\pi}$. Thus,

If X is a $\mathcal{N}(0, \sigma^2)$ randam variable, then $Y = aX^2$ represents a χ^2 random variable with one degree of freedom (i.e., $n = 1$).

Example 6.10 Let X and Y be independent Gaussian random variables, each distributed as $\mathcal{N}(0, \sigma^2)$.

1. If the random variables Z and W are defined as $Z = X + Y$ and $W = 2X - Y$, find the values of μ_Z, μ_W, σ_Z^2, and σ_W^2.
2. Find the joint probability density function of the random variables Z and W. What is the correlation coefficient between these two random variables?

Solution

1. We have $\mu_X = \mu_Y = 0$. Hence,

$$\mu_Z = E[Z] = E[X + Y] = E[X] + E[Y] = 0.$$

Similarly, $E[X^2] = E[Y^2] = \sigma^2$, and therefore,

$$\sigma_Z^2 = E[Z^2] - \{E[Z]\}^2$$

$$= E[Z^2]$$

$$= E[(X + Y)^2] = E[X^2] + E[Y^2] + E[XY] + E[YX]$$

$$= E[X^2] + E[Y^2] = 2\sigma^2$$

as X and Y are statistically independent. As X and Y are Gaussian random variables, so is $Z = X + Y$. Hence the PDF of Z is

$$f_Z(z) = \frac{1}{\sqrt{2\pi}\sigma_Z} \exp\left(-\frac{z^2}{2\sigma_Z^2}\right).$$

2. In this case, we have

$$\begin{aligned}
\mu_W &= 2\mu_X - \mu_Y = 0 \\
\sigma_W^2 &= E[W^2] - \{E[W]\}^2 \\
&= E[(2X - Y)^2] \\
&= E[4X^2] + E[Y^2] \\
&= 5\sigma^2
\end{aligned}$$

since, X and Y are statistically independent. As X and Y are Gaussian random variables, so is $W = 2X - Y$. Hence the PDF of W is

$$f_W(w) = \frac{1}{\sqrt{2\pi}\sigma_W} \exp\left(-\frac{w^2}{2\sigma_W^2}\right).$$

The joint probability density function of Z and W is given by

$$f_{Z,W}(z, w) = \frac{1}{2\pi\sigma_Z\sigma_W} \exp\left(-\left[\frac{z^2}{2\sigma_Z^2} + \frac{w^2}{2\sigma_W^2}\right]\right).$$

The covariance of Z and W is given by

$$\begin{aligned}
\text{COV}(Z, W) &\triangleq E[ZW] - \mu_Z\mu_W \\
&= E[ZW].
\end{aligned}$$

Now the correlation coefficient $\rho_{Z,W}$ is given by

$$\begin{aligned}
\rho_{Z,W} &\triangleq \frac{\text{COV}(Z, W)}{\sigma_Z\sigma_W} \\
&= \frac{E[ZW]}{\sigma_Z\sigma_W}.
\end{aligned}$$

$$\begin{aligned}\text{But, } E[ZW] &= E[(X+Y)(2X-Y)] \\ &= E[2X^2+2YX-XY-Y^2] \\ &= 2E[X^2]-E[Y^2]=\sigma^2.\end{aligned}$$

$$\text{Hence, } \rho_{Z,W} = \frac{\sigma^2}{\sqrt{2}\sigma \times \sqrt{5}\sigma} = \frac{1}{\sqrt{10}}.$$

Example 6.11 A discrete random variable has K equally likely values $0, a, 2a, \ldots, (K-1)a$. Find the mean second moment and standard deviation.

Solution

The mean of a discrete random variable is given by

$$\begin{aligned}E[X] &\triangleq \sum_{0}^{\infty} x\, P\{X=x\} \\ &= \frac{1}{K}\sum_{\mathrm{i=0}}^{\mathrm{K-1}} ai = \frac{(K-1)a}{2}.\end{aligned}$$

The second moment is given by

$$\begin{aligned}E[X^2] &\triangleq \sum_{0}^{\infty} x^2\, P\{X=x\} \\ &= \frac{1}{K}\sum_{\mathrm{i=0}}^{\mathrm{K-1}} (ai)^2 \\ &= \frac{a^2}{6}(K-1)(2K-1).\end{aligned}$$

Now the variance is given by $\sigma_{\mathrm{X}}^2 = E[X^2] - \mu_{\mathrm{X}}^2$:

$$\begin{aligned}\sigma_{\mathrm{X}}^2 &= \frac{a^2}{6}(K-1)(2K-1) - \frac{(K-1)^2a^2}{4} \\ &= \frac{a^2(K^2-1)}{12}\end{aligned}$$

$$\text{Hence, } \sigma_{\mathrm{X}} = a\sqrt{\frac{K^2-1}{12}}.$$

To evaluate the sum $\sum_{\mathrm{i=0}}^{\mathrm{K-1}} (ai)^2$, we can use the following MATLAB script.

```
syms K i mx2
mx2 = simple(a^2*symsum(i^2,0,(K-1)));
```

6.3 Random Processes or Stochastic Processes

A *random process* $X(t)$ is defined as an ensemble of time functions together with a probability rule that assigns a probability to any meaningful event associated with an observation of one of the sample functions of the random process. The random variable and random process can be distinguished as follows:

- For a random variable, the outcome of a random experiment is mapped into a number.
- For a random process, the outcome of a random experiment is mapped into a waveform that is a function of time.

Random processes are suitable for description of certain phenomena in signal transmission. Another situation where modelling by random processes proves to be useful is in the characterisation of information sources.

At any time instant the value of a random process constitutes a random variable.

6.4 Stationary Processes

If a random process is divided into a number of time intervals, and if the various sections of the process exhibit essentially the same statistical properties, then such a process is said to be *stationary*. Otherwise, it is said to be *non-stationary*. Generally speaking, a stationary process arises from a stable physical phenomenon that has evolved into a steady-state mode of behaviour, whereas a nonstationary process arises from an unstable phenomenon.

A random process $X(t)$, initiated at time $t = -\infty$, is strictly stationary if the joint distribution of any set of random variables obtained by observing the random process $X(t)$ is invariant with respect to the location of the origin $t = 0$.

We may say that two random processes $X(t)$ and $Y(t)$ are *jointly strictly stationary* if the joint finite-dimensional distributions of the two sets of random variables $X(t_1), \ldots, X(t_k)$ and $Y(t_1'), \ldots, Y(t_j')$ are invariant with respect to the origin $t = 0$ for all k and j and all choices of observation times $t_1, \ldots, t_k$ and $t_1', \ldots, t_j'$.

6.4.1 Cyclostationary processes

In *cyclostationary processes*, the statistical properties are not time dependent. A random process $X(t)$ with mean μ_X and autocorrelation function $R_X(t+\tau, t)$ is called cyclostationary if both the mean and the autocorrelation are periodic in t with some period T_0, i.e., if

$$\mu_X(t + T_0) = \mu_X(t) \tag{6.4.44}$$

and

$$R_X(t + \tau + T_0, t + T_0) = R_X(t + \tau, t) \tag{6.4.45}$$

for all t and τ.

6.4.2 Mean of a random process

Consider a strictly stationary random process $X(t)$. We define the *mean* of the process $X(t)$ as the expectation of the random variable obtained by observing the process at some time t, as given by

$$\mu_X(t) = E[X(t)] = \int_{-\infty}^{\infty} x f_{X(t)}(x)dx, \tag{6.4.46}$$

where $f_{X(t)}(x)$ is the first-order probability density function of the process. For a strictly stationary random process, $f_{X(t)}(x)$ is independent of time t. Consequently, *the mean of a strictly stationary process is a constant,* as shown by

$$\mu_X(t) = \mu_X, \quad \text{for all } t. \tag{6.4.47}$$

6.4.3 Auto- and cross-correlation functions

We define the *autocorrelation function* of the process $X(t)$ as the expectation of the product of two random variables, $X(t_1)$ and $X(t_2)$, obtained by observing the process $X(t)$ at time t_1 and t_2, respectively. We can write:

$$\begin{aligned} R_X(t_1, t_2) &= E[X(t_1)X(t_2)] \\ &= \int_{-\infty}^{\infty}\int_{-\infty}^{\infty} x_1 x_2 f_{X(t_1),X(t_2)}(x_1, x_2)dx_1 dx_2, \end{aligned} \tag{6.4.48}$$

where $f_{X(t_1),X(t_2)}(x_1, x_2)$ is the second-order probability density of the process. We can deduce that for a strictly stationary random process $f_{X(t_1),X(t_2)}(x_1, x_2)$ depends only on the difference between the observation times t_1 and t_2. Therefore,

The autocorrelation function of a strictly stationary process depends only on the time difference $t_1 - t_2$.

This can be stated as

$$R_X(t_1, t_2) = R_X(t_2 - t_1), \quad \text{for all } t_1 \text{ and } t_2. \tag{6.4.49}$$

The *cross-correlation function* of two random processes $X(t)$ and $Y(t)$ is defined as the expectation of the product of the two random variables, $X(t_1)$ and $Y(t_2)$. Thus, we have,

$$\begin{aligned} R_{XY}(t_1, t_2) &= E[X(t_1)Y(t_2)] \\ &= \int_{-\infty}^{\infty}\int_{-\infty}^{\infty} x_1 y_2 f_{X(t_1),Y(t_2)}(x_1, y_2)dx_1 dy_2. \end{aligned} \tag{6.4.50}$$

The correlation properties of two random processes $X(t)$ and $Y(t)$ may be displayed in matrix form as

$$R(t_1, t_2) = \begin{bmatrix} R_X(t_1, t_2) & R_{XY}(t_1, t_2) \\ R_{YX}(t_1, t_2) & R_Y(t_1, t_2) \end{bmatrix} \tag{6.4.51}$$

which is called the *correlation matrix* of the random processes $X(t)$ and $Y(t)$. If the two processes are each strictly stationary and jointly stationary, then this matrix can be written as

$$R(\tau) = \begin{bmatrix} R_X(\tau) & R_{XY}(\tau) \\ R_{YX}(\tau) & R_Y(\tau) \end{bmatrix}. \tag{6.4.52}$$

where $\tau = t_2 - t_1$.

6.4.4 Auto-covariance

The *auto-covariance* of a strictly stationary process $X(t)$ is given as

$$\begin{aligned} C_X(t_1, t_2) &= E[(X(t_1) - \mu_X)(X(t_2) - \mu_X)] \\ &= R_X(t_2 - t_1) - \mu_X^2. \end{aligned} \tag{6.4.53}$$

This equation states that, like the autocorrelation function, the auto-covariance function of strictly stationary process $X(t)$ depends only on the time difference $t_2 - t_1$. The following points are to be noted.

- The mean and autocorrelation function only provide a *partial description* of the distribution of a random process $X(t)$.
- The conditions of (6.4.47) and (6.4.49). involving the mean and autocorrelation function, respectively, are not sufficient to guarantee that the random process $X(t)$ is strictly stationary.

The class of random processes that satisfy (6.4.47) and (6.4.49) are in general called *second-order stationary, wide-sense stationary, or weakly stationary* processes.

6.4.5 Wide-sense stationary process

A random process satisfying the following properties is termed as a wide-sense stationary (WSS) process.

- A process is WSS if its expected power is finite, i.e., $E\left[X^2(t)\right] < 1$, its mean is constant, $E\left[X(t)\right] = \mu_X$, and its autocorrelation depends only on the time difference of the samples, $R_X(t1, t2) = R_X(\tau = t2 - t1)$.
- Intuitively, WSS processes are stationary in their first and second moments.
- Stationary processes are WSS but not vice versa.

6.4.6 Strict-sense stationary process

A random process is a strict-sense stationary (SSS) process if time shifts do not affect its probability. In other words, for all T and all sets of sample times

$$t_0,\ t_1,\ \ldots,\ t_n, P[X(t_0) \le x_0,$$

$$X(t_1) \le x_1,\ \ldots,\ X(t_n) \le x_n] = P[X(t_0+T) \le x_0,$$

$$X(t_1+T) \le x_1,\ \ldots,\ X(t_n+T) \le x_n].$$

Example 6.12 The sample function $x(t)$ of a random process $X(t)$ consists of a random sequence of binary symbols 1 and 0.

The MATLAB code to generate and plot the random binary as function m-file is given below. A typical plot is also appended.

```
% MATLAB function to generate and plot a random binary signal..
% nbits & nsamples are input arguments..
% x is the random binary signal.
function [x,bits]=random_binary(nbits,nsamples)
k=nbits*nsamples;
x=zeros(1,k);
bits=round(rand(1,nbits));
for m=1:nbits
  for n=1:nsamples
    index=(m-1)*nsamples+n;
    x(1,index)=(-1).^bits(m);
  end
end
plot([1:k],x); axis([1 k -1.2 1.2]);
```

Note that the function when called as $[x, \text{bits}] =$ `random_binary` $(32, 8192)$ will produce a waveform similar to the one given in Figure 6.20. It should be noted that the same function called with the same set of arguments, at some other instant of time, will produce a different result altogether!

Correlation coefficient

The *covariance* of two random processes $X(t)$ and $Y(t)$ are defined as

$$\begin{aligned} \mathrm{COV}(X,Y) &= E[(X-\mu_X)(Y-\mu_Y)] \\ &= E[X.Y - Y.\mu_X - X.\mu_Y + \mu_X.\mu_Y] \\ &= E[XY] - \mu_Y\mu_X - \mu_X\mu_Y + \mu_X\mu_Y \\ &= E[XY] - \mu_X\mu_Y. \end{aligned} \tag{6.4.54}$$

The normalised value of covariance is called *correlation coefficient* denoted by $\rho_{X,Y}$ is defined by

$$\rho_{X,Y} = \frac{\mathrm{COV}(X,Y)}{\sigma_X}\sigma_Y. \tag{6.4.55}$$

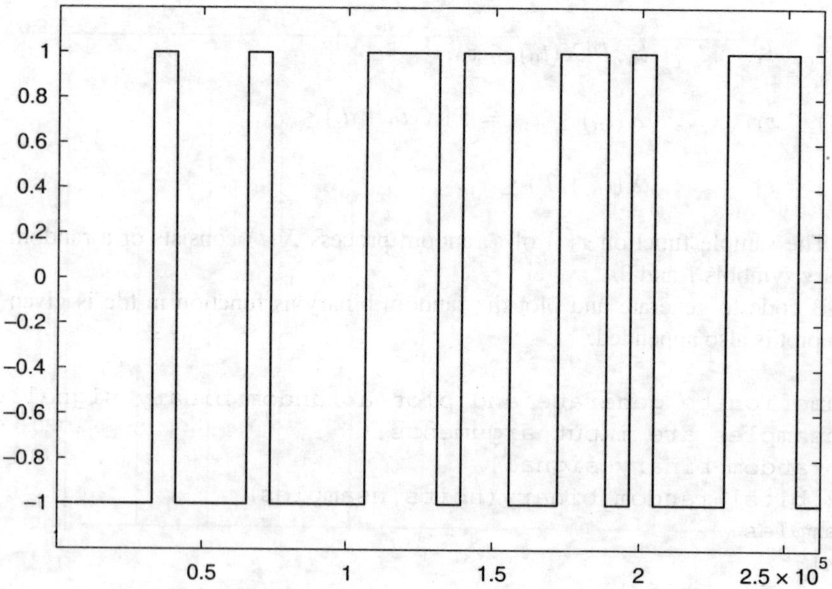

Figure 6.20 *A random binary signal*

Using the Cauchy–Schwarz inequality, it can be shown that $|\rho_{X,Y}| \leq 1$ and $\rho = \pm 1$ indicates a first-order relationship between X and Y, i.e., a relation of the form $Y = aX + b$. If X and Y are independent, then $\mathrm{COV}(X, Y) = \rho_{X,Y} = 0$. That is, independence is a lack of correlation.

Properties of autocorrelation function

The autocorrelation function of a stationary process $X(t)$ is defined as:

$$R_X(\tau) = E[X(t+\tau)X(t)], \quad \text{for all } t. \tag{6.4.56}$$

The following are the properties of the autocorrelation function:

1. The *mean square value* of the process may be obtained by putting $\tau = 0$ in (6.4.56).

$$R_X(0) = E[X^2(t)]. \tag{6.4.57}$$

2. The autocorrelation is an even function of τ, that is,

$$R_X(\tau) = R_X(-\tau). \tag{6.4.58}$$

3. The autocorrelation function has its maximum magnitude at $\tau = 0$, that is,

$$|R_X(\tau)| \leq R_X(0). \tag{6.4.59}$$

6.5 Ergodic Processes

The *expectations* or *ensemble averages* of a random process $X(t)$ are averages 'across the process'. As an example, the mean of a random process $X(t)$ at some fixed time t_k is the expectation of the random variable $X(t_k)$ that describes *all possible values* of the sample functions of the process observed at time $t = t_k$. We can also define *long-term sample averages*, or *time averages* that are averages 'along the process'. For a sample function $x(t)$ of a random process $X(t)$, with the observation interval as $-T \leq t \leq T$, the dc value of $x(t)$ is defined as by the time average

$$\mu_X(T) = \frac{1}{2T}\int_{-T}^{T} x(t)dt. \tag{6.5.60}$$

The time average $\mu_X(T)$ is a random variable as its value depends on the observation interval and the particular sample function of the random process $X(t)$ which is picked for use in (6.5.60). As the process $X(t)$ is assumed to be stationary, the mean of the time average $\mu_X(T)$ is given by

$$\begin{aligned} E[\mu_X(T)] &= E[\frac{1}{2T}\int_{-T}^{T} x(t)dt] \\ &= \frac{1}{2T}\int_{-T}^{T} E[x(t)]dt \\ &= \frac{1}{2T}\int_{-T}^{T} \mu_X dt = \mu_X, \end{aligned} \tag{6.5.61}$$

where μ_X is the mean of the random process $X(t)$. Therefore, the time average $\mu_X(T)$ represents an unbiased estimate of the ensemble-averaged mean μ_X. We can say that the process $X(t)$ is *ergodic in the mean* if two conditions are satisfied:

1. The time average $\mu_X(T)$ approaches the ensemble average μ_X in the limit as the observation interval T approaches ∞, that is,

$$\lim_{T\to\infty} \mu_X(T) = \mu_X. \tag{6.5.62}$$

2. The variance of $\mu_X(T)$, treated as a random variable, approaches 0 in the limit as the observation time T approaches ∞, that is,

$$\lim_{T\to\infty} \mathrm{VAR}[\mu_X(T)] = 0. \tag{6.5.63}$$

The *time-averaged autocorrelation function* is defined as

$$R_X(\tau, T) = \frac{1}{2T}\int_{-\infty}^{\infty} x(t+\tau)x(t)dt. \tag{6.5.64}$$

The process $X(t)$ is *ergodic in the autocorrelation function* if it satisfies the following two limiting conditions:

$$\lim_{T\to\infty} R_X(\tau, T) = R_X(\tau), \tag{6.5.65}$$

$$\lim_{T\to\infty} \text{VAR}[R_X(\tau, T)] = 0. \tag{6.5.66}$$

6.6 Response of Random Processes over an LTI System

Assuming that the random process $X(t)$ is stationary, its response, $Y(t)$, when passed through a *linear time invariant filter (system)* whose impulse response if $h(t)$, can be obtained by means of convolution integral as

$$Y(t) = \int_{-\infty}^{\infty} h(\tau)X(t-\tau)d\tau, \tag{6.6.67}$$

where τ is the integration variable. Hence the mean of $Y(t)$ is

$$\begin{aligned} \mu_Y(t) &= E[Y(t)] \\ &= E\left[\int_{-\infty}^{\infty} h(\tau)X(t-\tau)d\tau\right]. \end{aligned} \tag{6.6.68}$$

Given that the expectation $E[Y(t)]$ is finite for all t and the system is stable, we can interchange the order of expectation and integration in (6.6.68). So, we have

$$\begin{aligned} \mu_Y(t) &= \int_{-\infty}^{\infty} h(\tau)E[X(t-\tau)]d\tau \\ &= \int_{-\infty}^{\infty} \mu_X(t-\tau)d\tau. \end{aligned} \tag{6.6.69}$$

When the random process $X(t)$ is stationary, the mean $\mu_X(t)$ is a constant μ_X, so that we may simplify the preceding equation as follows:

$$\mu_Y = \mu_X \int_{-\infty}^{\infty} h(\tau)d\tau = \mu_X H(0), \tag{6.6.70}$$

where $H(0)$ is the dc response of the system. (6.6.70) states that the mean of the random process $Y(t)$ produced at the output of a LTI system in response to the input $X(t)$ is equal to the mean of $X(t)$ multiplied by the dc response of the system.

The autocorrelation function of the output random process $Y(t)$, by definition, is equal to

$$R_Y(t,u) = E[Y(t)Y(u)], \tag{6.6.71}$$

where t and u denote two values of the time at which the output process is observed. Using convolution integral we may write

$$R_Y(t,u) = E\left[\int_{-\infty}^{\infty} h(\tau_1)X(t-\tau_1)d\tau_1 \int_{-\infty}^{\infty} h(\tau_2)X(u-\tau_2)d\tau_2\right]. \tag{6.6.72}$$

If the mean square value $E[X^2(t)]$ is finite for all t and the system is stable, we can interchange the order of expectation and integration in (6.6.72). Then,

$$\begin{aligned} R_Y(t,u) &= \int_{-\infty}^{\infty} d\tau_1 h(\tau_1) \int_{-\infty}^{\infty} d\tau_2 h(\tau_2) E[X(t-\tau_1)X(u-\tau_2)] \\ &= \int_{-\infty}^{\infty} d\tau_1 h(\tau_1) \int_{-\infty}^{\infty} d\tau_2 h(\tau_2) R_X(t-\tau_1, u-\tau_2). \end{aligned} \tag{6.6.73}$$

When the input $X(t)$ is a stationary process, the autocorrelation function of $X(t)$ is only a function of the difference between the observation times $t - \tau_1$ and $u - \tau_2$. Thus putting $\tau = t - u$ in the preceding equation, we get

$$R_Y(t) = \int_{-\infty}^{\infty}\int_{-\infty}^{\infty} h(\tau_1)h(\tau_2)R_X(\tau - \tau_1 + \tau_2)d\tau_1\, d\tau_2. \tag{6.6.74}$$

Thus, we can conclude that

If the input to a stable LTI system is a stationary process, then the output of the system is also a stationary process.

As $R_Y(0) = E[Y^2(t)]$, it follows that the mean square value of the output random process $Y(t)$ is obtained by putting $\tau = 0$ in (6.6.74). Thus, we get the result

$$E[Y^2(t)] = \int_{-\infty}^{\infty}\int_{-\infty}^{\infty} h(\tau_1)h(\tau_2)R_X(\tau_2 - \tau_1)d\tau_1\, d\tau_2 \tag{6.6.75}$$

which is a constant.

6.7 Power Spectral Density

By definition, the impulse response of a LTI filter is equal to the inverse Fourier transform of the frequency response of the system. Thus,

$$h(\tau) = \int_{-\infty}^{\infty} H(f) \exp(j2\pi f\tau) df. \tag{6.7.76}$$

Substituting the corresponding expression for $h(\tau_1)$ in (6.6.75). we have

$$\begin{aligned} E[Y^2(t)] &= \int_{-\infty}^{\infty}\int_{-\infty}^{\infty}\left[\int_{-\infty}^{\infty} H(f) \exp(j2\pi f\tau) df\right] h(\tau_2) R_X(\tau_2 - \tau_1) d\tau_1\, d\tau_2 \\ &= \int_{-\infty}^{\infty} H(f)\, df \int_{-\infty}^{\infty} d\tau_2 h(\tau_2) \int_{-\infty}^{\infty} R_X(\tau_2 - \tau_1) \exp(j2\pi f\tau_1) d\tau_1. \end{aligned} \tag{6.7.77}$$

Now putting $\tau = \tau_2 - \tau_1$, we get

$$E[Y^2(t)] = \int_{-\infty}^{\infty} H(f)\, df \int_{-\infty}^{\infty} d\tau_2 h(\tau_2) \exp(j2\pi f\tau_2) \int_{-\infty}^{\infty} R_X(\tau) \exp(-j2\pi f\tau) d\tau. \tag{6.7.78}$$

The middle integral on the right side of (6.7.78) is $H^*(f)$, the complex conjugate of the frequency response of the filter. Now, we have

$$H(f).H^*(f) = |H(f)|^2. \tag{6.7.79}$$

Therefore, we can simplify (6.7.78) as

$$E[Y^2(t)] = \int_{-\infty}^{\infty} |H(f)|^2\, df \int_{-\infty}^{\infty} R_X(\tau) \exp(-j2\pi f\tau) d\tau \tag{6.7.80}$$

where $|H(f)|$ is the magnitude response of the filter. Now we define the *power spectral density (PSD)* or *power spectrum* of the stationary process $X(t)$ as

$$S_X(f) \triangleq \int_{-\infty}^{\infty} R_X(\tau) \exp(-j2\pi f\tau) d\tau. \tag{6.7.81}$$

Then, we can write (6.7.80) as

$$E[Y^2(t)] = \int_{-\infty}^{\infty} |H(f)|^2\, S_X(f)\, df. \tag{6.7.82}$$

Thus, we have an important relation:

The mean square value of the output of a stable LTI system in response to a stationary process is equal to the integral over all frequencies of the power spectral density of the input process multiplied by the squared of the magnitude response of the system.

6.7.1 Properties of power spectral density

The PSD $S_X(f)$ and the autocorrelation function $R_X(t)$ of a stationary random process $X(t)$ form a Fourier transform pair, i.e.,

$$S_X(f) = \int_{-\infty}^{\infty} R_X(\tau) \exp(-j2\pi f\tau) d\tau \tag{6.7.83}$$

$$R_X(\tau) = \int_{-\infty}^{\infty} S_X(f) \exp(j2\pi f\tau) df. \tag{6.7.84}$$

Einstein–Wiener–Khintchine theorem

Equations (6.7.83) and (6.7.84) are together known as the *Einstein–Wiener–Khintchine Theorem*. The properties of PSD are listed as follows.

1. The zero frequency value of the PSD of a stationary process equals the total area under the graph of the autocorrelation function, i.e.,

$$S_X(0) = \int_{-\infty}^{\infty} R_X(\tau) d\tau. \tag{6.7.85}$$

   This property follows directly from (6.7.83) by putting $f = 0$.

2. The mean square value of a stationary process equals the total area under the graph of PSD, i.e.,

$$E[X^2(t)] = \int_{-\infty}^{\infty} S_X(f) df. \tag{6.7.86}$$

   This property follows from (6.7.84) by putting $\tau = 0$ and noting that $R_X(0) = E[X^2(t)]$.

3. The PSD of a stationary random process is always *non-negative*, i.e.,

$$S_X(f) \geq 0, \text{ for all } f. \tag{6.7.87}$$

   This property is a direct consequence of the fact that the mean square value $E[X^2(t)]$ must always be non-negative.

4. The PSD of a real-valued random process is an even function of the frequency, i.e.,

$$S_X(-f) = S_X(f). \tag{6.7.88}$$

This property is derived by substituting $-f$ for f in (6.7.83) and noting that $R_X(-\tau) = R_X(\tau)$.

5. The normalised PSD has the properties associated with a probability density function. Consider the function

$$\rho_X(f) = \frac{S_X(f)}{\int_{-\infty}^{\infty} S_X(f)df}. \tag{6.7.89}$$

Now, we can show that $\rho_X(f) \geq 0$ for all f. Also, it can be shown that $\int_{-\infty}^{\infty} \rho_X(f)df = 1$. Thus the normalised PSD performs similar to a probability density function.

6.7.2 Relation among PSDs of input and output processes

Let $S_Y(f)$ denote the PSD of the output random process $Y(t)$ obtained by passing the random process $X(t)$ through a LTI filter with frequency response $H(f)$. Then, we have

$$\begin{aligned} S_Y(f) &= \int_{-\infty}^{\infty} R_Y(\tau) \exp(-j2\pi f\tau)d\tau \\ &= \int_{-\infty}^{\infty}\int_{-\infty}^{\infty}\int_{-\infty}^{\infty} h(\tau_1)h(\tau_2)R_X(\tau-\tau_1+\tau_2) \exp(-j2\pi f\tau)d\tau_1 d\tau_2 d\tau. \end{aligned} \tag{6.7.90}$$

Let $\tau_0 = \tau - \tau_1 + \tau_2$. Equation (6.7.90) reduces to

$$\begin{aligned} S_Y(f) &= \int_{-\infty}^{\infty}\int_{-\infty}^{\infty}\int_{-\infty}^{\infty} h(\tau_1)h(\tau_2)R_X(\tau_0) \exp(-j2\pi f\{\tau_0+\tau_1-\tau_2\})d\tau_1 d\tau_2 d\tau \\ &= \int_{-\infty}^{\infty} h(\tau_1) \exp(-j2\pi f\tau_1)d\tau_1 \int_{-\infty}^{\infty} h(\tau_2) \exp(j2\pi f\tau_2)d\tau_2 \\ &\quad \int_{-\infty}^{\infty} R_X(\tau_0) \exp(-j2\pi f\tau)d\tau_0 \\ &= H(f)H^*(f)S_X(f) = |H(f)|^2 S_X(f). \end{aligned} \tag{6.7.91}$$

Thus, we have:

The PSD of the output random process $Y(t)$ equals the PSD of the input process $X(t)$ multiplied by the squared magnitude response of the of the LTI filter.

6.8 Cross Spectral Density

Cross spectral densities provide a measure of frequency inter-relationship between two random processes. Let $X(t)$ and $Y(t)$ be two jointly stationary random processes with their cross correlation functions denoted by $R_{\mathrm{XY}}(\tau)$ and $R_{\mathrm{YX}}(\tau)$. Now by definition, the *cross spectral densities* $S_{\mathrm{XY}}(f)$ and $S_{\mathrm{YX}}(f)$ are the Fourier transforms of the respective cross correlation functions, i.e.,

$$S_{\mathrm{XY}}(f) = \int_{-\infty}^{\infty} R_{\mathrm{XY}}(\tau) \exp(-j2\pi f\tau)d\tau, \tag{6.8.92}$$

$$S_{\mathrm{YX}}(f) = \int_{-\infty}^{\infty} R_{\mathrm{YX}}(\tau) \exp(-j2\pi f\tau)d\tau. \tag{6.8.93}$$

As the cross correlation functions and cross spectral densities form Fourier transform pairs, we can also write the inverse relationships as

$$R_{\mathrm{XY}}(\tau) = \int_{-\infty}^{\infty} S_{\mathrm{XY}}(f) \exp(j2\pi f\tau)df, \tag{6.8.94}$$

$$R_{\mathrm{YX}}(\tau) = \int_{-\infty}^{\infty} S_{\mathrm{YX}}(f) \exp(j2\pi f\tau)df. \tag{6.8.95}$$

As $R_{\mathrm{XY}}(\tau) = R_{\mathrm{YX}}(-\tau)$, we can show that

$$S_{\mathrm{XY}}(f) = S_{\mathrm{YX}}(-f) = S^*_{\mathrm{YX}}(f). \tag{6.8.96}$$

Example 6.13: Let two random processes $X(t)$ and $Y(t)$ have zero mean and they be individually stationary. Consider a random process $Z(t)$, which is the sum $X(t) + Y(t)$. Obtain the PSD of $Z(t)$.

Solution
The autocorrelation function of $Z(t)$ is given by

$$\begin{aligned} R_{\mathrm{Z}}(t,u) &= E[Z(t)Z(u)] \\ &= E[(X(t)+Y(t))(X(u)+Y(u))] \\ &= E[X(t)X(u)] + E[X(t)Y(u)] + E[Y(t)X(u)] + E[Y(t)Y(u)] \\ &= R_{\mathrm{X}}(t,u) + R_{\mathrm{XY}}(t,u) + R_{\mathrm{YX}}(t,u) + R_{\mathrm{Y}}(t,u). \end{aligned}$$

Putting $\tau = t - u$, we can also write

$$R_Z(\tau) = R_X(\tau) + R_{XY}(\tau) + R_{YX}(\tau) + R_Y(\tau) \tag{6.8.97}$$

when the random processes $X(t)$ and $Y(t)$ are also jointly stationary. Now taking the Fourier transforms of both sides of (6.8.97),

$$S_Z(f) = S_X(f) + S_{XY}(f) + S_{YX}(f) + S_Y(f). \tag{6.8.98}$$

When the stationary processes $X(t)$ and $Y(t)$ are uncorrelated, the cross spectral densities $S_{XY}(f)$ and $S_{YX}(f)$ are zero, and therefore (6.8.98) reduces to

$$S_Z(f) = S_X(f) + S_Y(f). \tag{6.8.99}$$

6.9 Gaussian Process

Let there be a random process $X(t)$ observed between the interval $t = 0$ and $t = T$. If we weight the random process $X(t)$ by a function $\psi(t)$ and then integrate the product $\psi(t)X(t)$ over the observation interval, we obtain a random variable Y defined by

$$Y = \int_0^T \psi(t)X(t)\,dt. \tag{6.9.100}$$

We refer to Y as a *functional* of $X(t)$. If in (6.9.100) the weighting function $\psi(t)$ is such that the mean square value of the random variable Y is finite, and if random variable Y is Gaussian-distributed for every $\psi(t)$ in this class of functions, then the process $X(t)$ is said to be a *Gaussian random process.* In other words, the process $X(t)$ is Gaussian if every linear functional of $X(t)$ is a Gaussian random variable. The random variable Y has a Gaussian distribution if its probability density function has the form

$$f_Y(y) = \frac{1}{\sqrt{2\pi}\sigma_Y} \exp\left[-\frac{(y-\mu_Y)^2}{2\sigma_Y^2}\right], \tag{6.9.101}$$

where μ_Y is the mean and σ_Y^2 is the variance of the random variable Y.[2] A Gaussian distribution with mean μ and variance σ^2 is referred to as $\mathcal{N}(\mu, \sigma^2)$.

6.9.1 Properties of Gaussian process

A Gaussian process has the following properties:

1. If a Gaussian process is applied to a LTI filter, the output produced by the filter will also be a Gaussian process.
2. A set of random variables or samples $X(t_1), X(t_2), \ldots, X(t_n)$ obtained by observing a Gaussian random variable $X(t)$ at times $t_1, t_2, \ldots, t_n$ is jointly Gaussian for any n, with their *n-fold joint probability density function* being completely determined by specifying the set of means

$$\mu_X(t_i) = E[X(t_1)], \quad t = 1, 2, \ldots, n$$

[2]Note that variance can never be negative. (Why?)

and the set of covariance functions

$$C_X(t_k, t_i) = E[(X(t_k) - \mu_{X(t_k)})(X(t_i) - \mu_{X(t_i)})], \quad k, i = 1, 2, \ldots, n.$$

3. If a Gaussian process is stationary, then the process is also strictly stationary.
4. If the random variables obtained by sampling a Gaussian random process are uncorrelated, then these random variables are statistically independent.

6.10 The Central Limit Theorem

The *central limit theorem* states that the probability distribution of a random process obtained by summing a set of *independent and identically distributed (i.i.d.)* random variables and normalising the result approaches a normalised Gaussian distribution $\mathcal{N}(0, 1)$ in the limit as the number of random variables N approaches infinity.

Let X_i be i.i.d. random variables. Let $Y = \sum_{i=1}^{n} X_i$ and $Z = \left(\frac{Y - \mu_Y}{\sigma_Y}\right)$. Now the central limit theorem states that the distribution of Z as $n \to \infty$ converges to a Gaussian random variable, $\mathcal{N}(0, 1)$.

The central limit theorem provides the mathematical justification for using a Gaussian process as a model for a large number of different physical phenomena in which the observed random variable, at a particular instant of time, is the result of a large number of individual random events.

Exercises

1. A random experiment consists of drawing a ball from an urn containing four red balls numbered 1, 2, 3, 4 and three black balls numbered 1, 2, 3. Precisely state what outcomes are contained in the following events:

   (a) E_1= The number on the ball is odd. (Answer: $E_1 = \{r1, r3, b1, b3\}$)

   (b) E_2= The ball is red and its number is greater than 2.(Answer: $E_2 = \{r3, r4\}$)

   (c) E_3= The number on the ball is less than 2. (Answer: $E_3 = \{r1, b1\}$)

   (d) $E_4 = E_1 \cap E_2$ (Answer: $E_4 = \{r3\}$)

2. If in Problem 1 all the balls are equally likely to be drawn, obtain the respective probabilities. (Answer: $P(E_1) = \frac{4}{7}, \;\; P(E_2) = \frac{2}{7}, \;\; P(E_3) = \frac{2}{7}, \;\; P(E_4) = \frac{1}{7}$.)
3. A biased coin is tossed such that $p(H) = (1 + \epsilon)/2$ with $0 < |\epsilon| < 1$. Show that the probability of a match in two independent tosses is greater than 1/2.
4. A coin is flipped three times and the random variable X denotes the total number of heads that turn up. The probability of a head in one flip of the coin is p.

   (a) What values can the random variable X take? (Answer: 0, 1, 2, 3)

   (b) What is the probability mass function of X? (Answer: $\binom{3}{k} p^k (1 - p)^{3-k}, \quad k = 0, 1, 2, 3$)

   (c) Derive the cumulative distribution function of X. (Answer: $\sum_{0}^{3} \binom{3}{k} p^k (1 - p)^{3-k}$)

(d) What is the probability that X exceeds 1? (Answer: $\sum_{2}^{3} \binom{3}{k} p^k (1-p)^{3-k} = 3p^2 (1-p) + p^3 = 3p^2 - 2p^3$)

5. A discrete random variable has two possible values, a and b. Find the mean, second moment and standard deviation in terms of $p = P(X = a)$. (Answer: $\mu_X = pa + (1-p)b$, $E[X^2] = pa^2 + (1-p)b^2$ $\sigma_X = (a-b)\sqrt{p(1-p)}$).
6. Find the mean, second moment and standard deviation of X where $f_X(x) = a \exp(-ax)u(x)$ with $a > 0$. (Answer: mean $= \frac{1}{a}$, $E[X^2] = \frac{2}{a^2}$, $\sigma_X = \frac{1}{a}$).
7. Obtain the mean and variance of a Poisson distributed random variable.
8. The noise voltage in an electric circuit can be modelled as a Gaussian random variable with zero mean and variance equal to 10^{-8}.

   (a) What is the probability that the value of the noise exceeds 10^{-4}? What is the probability that it exceeds 4×10^{-4}?

   (b) Given that the value of noise is positive, what is the probability that it exceeds 10^{-4}?

9. X is an $\mathcal{N}(0, \sigma^2)$ random variable. This randam variable is passed through a system whose input–output relation is given by $y = ax^2$. Find the PDF of the output variable Y.
10. The random variable Φ is uniformly distributed on the interval $[-\frac{\pi}{2}, \frac{\pi}{2}]$. Find the PDF of $X = \tan\,\Phi$. Find the mean and variance of X.
11. Let $Y = \alpha X + \beta$. Show that $\sigma_Y = |\alpha|\sigma_X$.
12. Show that for the binomial random variable, the mean is given by np and the variance is given by $np(1-p)$.
13. Let X and Y be independent random variables with

$$f_X(x) = \begin{cases} \alpha e^{-\alpha x}, & x > 0 \\ 0, & \text{otherwise} \end{cases}$$

and

$$f_Y(Y) = \begin{cases} \beta e^{-\beta y}, & y > 0 \\ 0, & \text{otherwise,} \end{cases}$$

where α and β are assumed to be positive constants. Find the PDF of $X + Y$.

14. The random variable X is uniform in the interval $[0, 1]$. Find the PDF of the random variable $Y = -\ln X$.
15. A coin whose probability of a head is $\frac{1}{4}$, is flipped 2000 times. Using the *central limit theorem* find the probability that the total number of heads lies between 480 and 520.
16. Let the random process $X(t)$ be defined by $X(t) = A + Bt$, where A and B are independent random variables each uniformly distributed on $[-1, 1]$. Find $m_X(t)$ and $R_X(t_1, t_2)$.
17. The stationary random process $X(t)$ has a PSD denoted by $S_X(f)$.

   (a) What is the PSD of $Y(t) = X(t) - X(t-T)$?

(b) What is the PSD of $Z(t) = X'(t) - X(t-T)$?

(c) What is the PSD of $W(t) = Y(t) + Z(t)$?

18. The random variables X and Y are $\mathcal{N}(0, 4)$ and independent. Find $F_Z(z)$ if (a) $Z = 2X + 3Y$, (b) $Z = \frac{X}{Y}$.
19. Show that the *convolution* of two normal densities is normal.
20. Find the probability density function of $Z = X + 3Y$ when X and Y are Gaussian random variables with $\mu_X = 6,\ \ \mu_Y = -2,\ \ \sigma_X = \sigma_Y = 4$, and $E[XY] = -22$.
21. Generate 100 random numbers as an array $\mathcal{N}$, normally distributed with zero mean and a variance of 4, using MATLAB. Verify that $\lim_{i} \rightarrow \infty$ mean $(N_i) = 0$.
22. Compute the autocorrelation function of the sequence obtained in the above problem. (Hint: Use the MATLAB *xcorr*(.) function.)

Chapter 7

The Effect of Noise on Carrier Wave Modulation

In chapters 3, 4 and 5 we discussed various analog modulation systems. In this chapter a topic of complementary nature, i.e. noise considerations in the design and analysis of analog modulation systems is discussed. We have seen that amplitude modulation (AM) is more prone to noise compared to frequency modulation (FM) and phase modulation (PM). Here we attempt an analytical study of noise in AM and angle modulation systems. We begin with the concept of normalised transmission bandwidth.

7.1 Normalised Transmission Bandwidth

The normalised transmission bandwidth is defined by

$$\eta = \frac{B_T}{W}, \tag{7.1.1}$$

where B_T is the transmission bandwidth of the modulated signal and W is the message bandwidth. By message bandwidth we mean the band of frequencies occupied by the message signal spectrum. The normalised transmission bandwidth points towards the amount of spreading of the signal spectrum while it is being transmitted. We will now compute the values of η for modulation schemes of AM, DSB-SC, SSB and FM.

Normalised transmission bandwidth of AM

For AM, the transmission bandwidth, B_T, is $2W$. Hence the value for η will be

$$\eta = \frac{B_T}{W} = \frac{2W}{W} = 2. \tag{7.1.2}$$

Normalised transmission bandwidth of DSB-SC

It can be shown that the normalised transmission bandwidth of DSB-SC is also equal to 2.

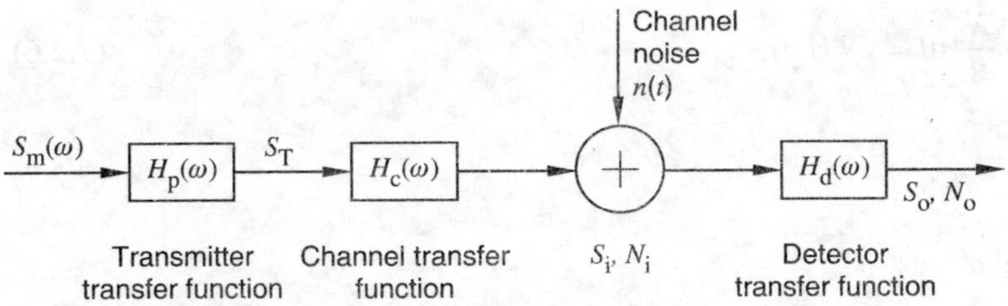

Figure 7.1 *Noise model for a baseband system*

Normalised transmission bandwidth of SSB

For SSB, the transmission bandwidth B_T is W as there is only one sideband to be counted. Hence the value for η will be

$$\eta = \frac{B_T}{W} = \frac{W}{W} = 1. \tag{7.1.3}$$

Normalised transmission bandwidth of FM

In the case of wideband FM, B_T is given by $2(\beta+2)W$. Hence the value for η is $2(\beta+2) \approx 2\beta$ for large values of β.

7.2 Noise in Baseband Systems

In baseband systems, the signal is transmitted directly without any modulation. This mode of communication is suitable over a pair of wires, optical fibre, or coaxial cables. It is mainly used in short-haul links. Baseband systems provide a basis against which other systems may be compared.

For a baseband system, the transmitter and receiver are ideal bandpass filters. See Figure 7.1 for details. The lowpass filter $H_{\text{p}}(\omega)$ at the transmitter limits the input signal spectrum to a given bandwidth. The lowpass filter $H_{\text{d}}(\omega)$ at the receiver eliminates the out-of-band noise and other channel interference.

The baseband signal is assumed to be a zero mean, wide-sense stationary random process bandlimited to B Hz. For a distortionless channel,

$$S_{\text{o}} = S_{\text{i}} \tag{7.2.4}$$

and

$$N_o = 2\int_0^B S_{\text{n}}(\omega)\, df, \tag{7.2.5}$$

where $S_{\text{n}}(\omega)$ is the power spectral density (PSD) of the channel noise. For the case of white noise, $S_{\text{n}}(\omega) = \frac{N}{2}$ and

$$N_o = 2\int_0^B \frac{N}{2}\,df = NB \tag{7.2.6}$$

and $$\frac{S_o}{N_o} = \frac{S_i}{NB}. \tag{7.2.7}$$

We define a parameter γ as

$$\gamma = \frac{S_i}{NB} = \frac{S_o}{N_o}. \tag{7.2.8}$$

For voice signals, a signal-to-noise ratio (SNR) of 5 to 10 dB at the receiver implies that the signal is barely intelligible. Telephone-quality signals have an SNR of 25 to 35 dB whereas television signals require an SNR of 45 to 55 dB.

7.3 Noise in AM Receivers

For AM receivers that carry analog-type message signals, SNR is the most commonly used measure of performance. All the noise generated within the receiver can be referred to the receiver input, which makes its easy to compare the receiver noise, antenna noise, and received signal. The antenna and receiver powers can be added, and so the receiving system can be modelled as shown in Figure 7.2.

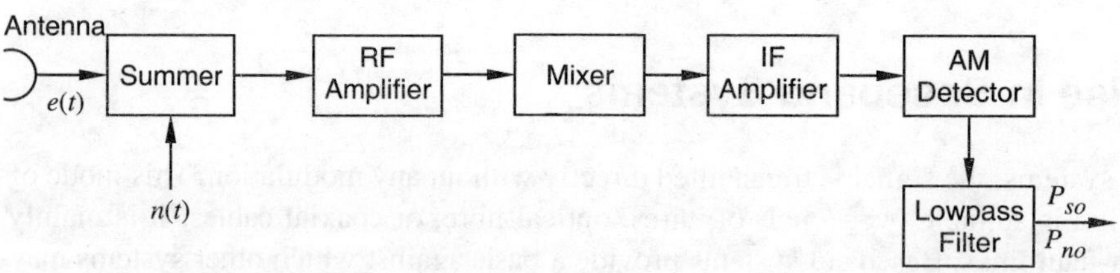

Figure 7.2 *Noise model of AM receiver*

For determining SNR, the bandwidth of the system is also required. From Figure 7.2, it is clear that between the antenna and the detector stage there is the bandwidth of the radiofrequency (RF) stages, followed by the bandwidth of the intermediate frequency (IF) stages. Normally, the bandwidth of the IF stages is very much smaller than the RF stages, and hence the IF bandwidth will decide the noise that reaches the detector. From the detector onwards the bandwidth will be that of the baseband. We denote the baseband bandwidth by W and the IF bandwidth by B_{IF}. For AM systems, it may be assumed that $B_{IF} \cong 2W$.

The noise output from a bandpass system is given by

$$n_{IF}(t) = n_I(t)\cos 2\pi f_{IF}\,t - n_Q(t)\sin 2\pi f_{IF}\,t, \tag{7.3.9}$$

where, $n_I(t)$ is the in-phase component and $n_Q(t)$ is the quadrature component of noise signal.

The modulated carrier can be represented by

$$\begin{aligned} e(t) &= E_{\text{c max}}(1 + m \cos 2\pi f_{\text{m}} t) \cos 2\pi f_{\text{IF}} t \\ &= A_{\text{c}}(t) \cos 2\pi f_{\text{IF}} t. \end{aligned} \tag{7.3.10}$$

The input to the detector is therefore

$$\begin{aligned} e_{\text{det}}(t) &= e(t) + n_{\text{IF}}(t) \\ &= (A_{\text{c}}(t) + n_{\text{I}}(t)) \cos 2\pi f_{\text{IF}} t - n_{\text{Q}}(t) \sin 2\pi f_{\text{IF}} t. \end{aligned} \tag{7.3.11}$$

Now, let $B(t) = \sqrt{[A_{\text{c}}(t) + n_{\text{I}}(t)]^2 + n_{\text{Q}}^2(t)}$, $\cos A = \frac{A_{\text{c}}(t)+n_{\text{I}}(t)}{B(t)}$ and $\sin A = \frac{n_{\text{Q}}(t)}{B(t)}$. Applying the trigonometric identity, we have,

$$e_{\text{det}}(t) = B(t).\ \cos\left(2\pi f_{\text{IF}} t + \psi(t)\right), \tag{7.3.12}$$

where, $\psi(t) = \tan^{-1}\left[\frac{n_{\text{Q}}(t)}{A_{\text{c}}(t)+n_{\text{I}}(t)}\right]$.

The AM envelope detector recovers the envelope of the waveform, and therefore the waveform needs to be expressed as shown in (7.3.12). The phase angle can be ignored and the amplitude term B can be expressed as

$$\begin{aligned} B(t) &= \sqrt{A_{\text{c}}^2(t) + 2A_{\text{c}}(t)n_{\text{I}}(t) + n_{\text{I}}^2(t) + n_{\text{Q}}^2(t)} \\ &\cong \sqrt{A_{\text{c}}^2(t) + 2A_{\text{c}}(t)n_{\text{I}}(t)} \end{aligned} \tag{7.3.13}$$

as the AM carrier amplitude is much higher than the noise. The expression for $B(t)$ can be further simplified as

$$\begin{aligned} B(t) &= \sqrt{A_{\text{c}}^2(t) + 2A_{\text{c}}(t)n_{\text{I}}(t)} \\ &= A_{\text{c}}(t)\left(1 + 2\frac{n_{\text{I}}(t)}{A_{\text{c}}(t)}\right)^{1/2} \\ &\cong A_{\text{c}}(t) + n_{\text{I}}(t) \qquad (7.3.14) \\ &= E_{\text{c max}} + mE_{\text{c max}} \cos 2\pi f_{\text{m}} t + n_{\text{I}}(t). \qquad (7.3.15) \end{aligned}$$

The envelope consists of a dc term, the modulating signal voltage and the noise voltage, $n_{\text{I}}(t)$. The available noise power output is given by

$$P_{\text{no}} = 2kT_{\text{s}} W. \tag{7.3.16}$$

The peak signal voltage at the output is $mE_{\text{c max}}$, and hence the RMS voltage is mE_{c}, where E_{c} is the rms voltage of the unmodulated carrier at the receiver. Therefore, the available signal power at the output is

$$P_{so} = \frac{m^2 E_c^2}{4R_{out}}. \quad (7.3.17)$$

The output SNR is

$$\left(\frac{S}{N}\right)_o = \frac{P_{so}}{P_{no}} = \frac{m^2 E_c^2}{8R_{out}kT_s W}. \quad (7.3.18)$$

The reference noise power at the detector is given by

$$P_{n\,REF} = kT_s\, W. \quad (7.3.19)$$

The available signal power from a source with internal resistance R_s is

$$P_R = \frac{E_c^2}{4R_s}\left(1 + \frac{m^2}{2}\right). \quad (7.3.20)$$

Therefore, the SNR is

$$\left(\frac{S}{N}\right)_{REF} = \frac{P_R}{P_{n\,REF}} \quad (7.3.21)$$

$$= \frac{E_c^2(1 + m^2/2)}{4R_s kT_s W}. \quad (7.3.22)$$

A *figure of merit* used is the ratio of the above two SNRs.

$$\text{FOM}_{AM} = \frac{(S/N)_o}{(S/N)_{REF}}$$

$$= \frac{m^2}{2 + m^2} \cdot \frac{R_s}{R_{out}}. \quad (7.3.23)$$

The higher the figure of merit, the better the system. Normally, $R_{out} \cong R_s$ and hence the highest value of FOM is $\frac{1}{3}$, achieved at 100% modulation depth.

7.4 Noise in Coherent DSB-SC Receivers

In the double sideband suppressed carrier (DSB-SC) case, the received signal is of the form

$$e(t) = E_{max} \cos \omega_m t \cos \omega_{IF} t, \quad (7.4.24)$$

where E_{max} is the peak value of the received signal. The input to the detector is therefore

$$
\begin{aligned}
e_{\text{det}}(t) &= e(t) + n(t) \\
&= (E_{\text{max}} \cos \omega_{\text{m}} t + n_{\text{I}}(t)) \cos \omega_{\text{IF}} t - n_{\text{Q}}(t) \sin \omega_{\text{IF}} t \\
&= A(t) \cos \omega_{\text{IF}} t - n_{\text{Q}}(t) \sin \omega_{\text{IF}} t, \qquad (7.4.25)
\end{aligned}
$$

where $A(t) = E_{\text{max}} \cos \omega_{\text{m}} t + n_{\text{I}}(t)$. Demodulation of DSB-SC makes use of a balanced mixer. The output of the balanced demodulator is

$$
e_{\text{out}}(t) = \beta\, e_{\text{det}}(t) \cos \omega_{\text{IF}} t, \qquad (7.4.26)
$$

where β is a constant of the multiplier circuit. Substituting the value of e_{det} from (7.4.25),

$$
\begin{aligned}
e_{\text{out}}(t) &= \frac{\beta}{2} A(t) + \frac{\beta}{2} A(t) \cos 2\omega_{\text{IF}} t - \frac{\beta}{2} n_{\text{Q}}(t) \sin 2\omega_{\text{IF}} t \qquad (7.4.27) \\
&= \frac{\beta}{2} (E_{\text{max}} \cos \omega_{\text{m}} t + n_I(t)) + \text{ higher-order terms}. \qquad (7.4.28)
\end{aligned}
$$

The lowpass filter following the balanced demodulator removes the high-frequency terms, leaving the baseband output.

$$
e_{\text{B}}(t) = \frac{\beta}{2} (E_{\text{max}} \cos \omega_{\text{m}} t + n_{\text{I}}(t)). \qquad (7.4.29)
$$

The $\frac{\beta}{2}$ factor is common to both signal and noise and can be ignored. Equation (7.3.18) is applicable to DSB-SC case too, with $\frac{E_{\text{max}}}{\sqrt{2}}$ replacing $m\, E_{\text{c}}$.

$$
\begin{aligned}
\left(\frac{S}{N}\right)_{\text{o}} &= \frac{(E_{\text{max}}/\sqrt{2})^2}{8R_{\text{out}}\, k\, T_{\text{s}}\, W} \\
&= \frac{E^2_{\text{max}}}{16R_{\text{out}}} k\, T_{\text{s}}\, W.
\end{aligned}
$$

The reference noise power is $P_{\text{n REF}} = kT_{\text{s}}\, W$. The RMS value of the received DSB-SC signal $E_{\text{max}} \cos \omega_{\text{m}} t \cos \omega_{\text{IF}} t$ is $\frac{E_{\text{max}}}{2}$. Hence, the available signal power at the input is

$$
P_{\text{R}} = \frac{(E_{\text{max}}/2)^2}{4R_{\text{s}}} = \frac{E^2_{\text{max}}}{16R_{\text{s}}}. \qquad (7.4.30)
$$

The reference SNR is therefore

$$
\left(\frac{S}{N}\right)_{\text{o}} = \frac{E^2_{\text{max}}}{16R_{\text{s}}\, k\, T_{\text{s}}\, W}. \qquad (7.4.31)
$$

The *figure of merit* is therefore

$$\mathrm{FOM}_{\mathrm{DSB-SC}} = \frac{(S/N)_{\mathrm{o}}}{(S/N)_{\mathrm{REF}}} = \frac{R_{\mathrm{s}}}{R_{\mathrm{out}}}. \tag{7.4.32}$$

For $R_{\mathrm{s}} = R_{\mathrm{out}}$, the $\mathrm{FOM}_{\mathrm{DSB-SC}}$ is unity, which is three time better than that of AM.

7.5 Noise in FM Receivers

In FM systems amplitude limiters help to reduce the impulse-type noise. Therefore, FM has an advantage over AM. Another advantage arises from the nature of the noise modulation process. The noise at the receiver input cannot directly frequency-modulate the incoming carrier since its frequency is fixed by the remote transmitter which may, in fact be crystal controlled.

An important advantage of the FM reception is that an improvement in SNR can be achieved by increasing the frequency deviation. This requires an increase in bandwidth, but there is at least an option of trading bandwidth for SNR.

Since the power spectral density (PSD) is kT_{s}, the available noise power at the detector input for a bandwidth W is $P_{\mathrm{nREF}} = kT_{\mathrm{s}}\,W$. The RMS signal voltage is $E_{\mathrm{c}} = E_{\mathrm{c\,max}}/\sqrt{2}$, and hence the signal power at the detector input is $P_{\mathrm{R}} = \frac{E_{\mathrm{c\,max}}^2}{8R_{\mathrm{s}}}$. Hence the reference SNR for the FM system is

$$\left(\frac{S}{N}\right)_{\mathrm{REF}} = \frac{P_{\mathrm{R}}}{P_{\mathrm{n\,REF}}} = \frac{E_{\mathrm{c\,max}}^2}{8R_{\mathrm{s}}kT_{\mathrm{s}}\,W}. \tag{7.5.33}$$

It should be noted that though the reference SNR is decided at the input side of the detector, the bandwidth W is decided by the lowpass filter at the output side.

Consider the signal output from the band-pass filter in the absence of noise when a sinusoidally modulated carrier is received. The instantaneous frequency is given by

$$f_{\mathrm{i}} = f_{\mathrm{IF}} + \Delta f\ \cos\,\omega_{\mathrm{m}}\,t. \tag{7.5.34}$$

The FM discriminator converts this to a signal output given by

$$v_{\mathrm{m}}(t) = \beta\Delta f\ \cos\,\omega_{\mathrm{m}}\,t, \tag{7.5.35}$$

where β, the frequency-to-voltage coefficient of the discriminator, is a constant. The rms voltage at the output is $E_{\mathrm{m}} = \frac{\beta\Delta f}{\sqrt{2}}$, and hence the available power output is

$$P_{\mathrm{so}} = \frac{(\beta\Delta f)^2}{8R_{\mathrm{out}}}. \tag{7.5.36}$$

The noise voltage from a bandpass system is given by (7.3.9), which can be written as

$$n_{\mathrm{IF}}(t) = n_{\mathrm{I}}(t)\ \cos\,\omega_{\mathrm{IF}}\,t\ -\ n_{\mathrm{Q}}(t)\ \sin\,\omega_{\mathrm{IF}}\,t. \tag{7.5.37}$$

The incoming carrier in this case is frequency modulated and can be written as

$$e(t) = E_{\mathrm{c\,max}}\ \cos\,(\omega_{\mathrm{IF}}\,t + \phi_{\mathrm{m}}(t)), \tag{7.5.38}$$

where $\phi_{\text{m}}(t)$ is the equivalent PM produced by the signal. The noise terms in the output, resulting from the interaction of $\phi_{\text{m}}(t)$ and the noise modulation, lie outside the baseband. Therefore the noise analysis can be done assuming that an unmodulated carrier is received. The input to the detector is

$$e_{\text{det}} = e_{\text{c}}(t) + n_{\text{IF}}(t) = (E_{\text{c max}} + n_{\text{I}}(t)) \cos \omega_{\text{IF}} t - n_{\text{Q}}(t) \sin \omega_{\text{IF}} t. \tag{7.5.39}$$

We can express this waveform in an equivalent form as $e_{\text{det}}(t) = B(t) \cos (\omega_{\text{IF}} t + \psi(t))$, where

$$B(t) = \sqrt{(E_{\text{c max}} + n_{\text{I}}(t))^2 + n_{\text{Q}}^2(t)} \tag{7.5.40}$$

$$\psi(t) = \tan^{-1}\left(\frac{n_{\text{Q}}(t)}{E_{\text{c max}} + n_{\text{I}}(t)}\right). \tag{7.5.41}$$

As the carrier amplitude is much greater than the noise voltage, the phase angle becomes

$$\psi(t) \cong \tan^{-1}\left(\frac{n_{\text{Q}}(t)}{E_{\text{c max}}}\right) \cong \frac{n_{\text{Q}}(t)}{E_{\text{c max}}}. \tag{7.5.42}$$

The total carrier angle is $\theta(t) = \omega_{\text{IF}} t + \psi(t)$ and therefore the equivalent frequency modulation is

$$f_{\text{ieq}}(t) = f_{\text{IF}} + \frac{1}{2\pi}\frac{d\psi(t)}{dt} = f_{\text{IF}} + \frac{\dot{n}_{\text{Q}}(t)}{2\pi E_{\text{c max}}}. \tag{7.5.43}$$

The noise voltage at output is therefore

$$v_{\text{n}}(t) = \frac{\beta \dot{n}_{\text{Q}}(t)}{2\pi E_{\text{c max}}}. \tag{7.5.44}$$

Fourier analysis shows that if a voltage waveform $v(t)$ has a PSD $G(f)$, then the PSD for $\dot{v}(t)$ is $\omega^2 G(f)$. The PSD for $n_{\text{Q}}(t)$ is $2kT_{\text{s}}$, and hence the PSD for $\dot{n}_{\text{Q}}(t)$ is $\omega^2\, 2kT_{\text{s}}$. Therefore, we can show that the PSD of $v_n(t)$ is

$$G_{\text{v}}(f) = \frac{\beta^2 f^2 2kT_{\text{s}}}{E_{\text{c max}}^2} \tag{7.5.45}$$

From the definition of PSD, the noise power at the output is

$$P_{\text{no}} = \int_0^W G_{\text{v}}(f) df = \frac{W^3 \beta^2 2kT_{\text{s}}}{3E_{\text{c}}^2 \text{max}}. \tag{7.5.46}$$

Therefore, the SNR at the output is

$$\left(\frac{S}{N}\right)_{\text{o}} = \frac{P_{\text{so}}}{P_{\text{no}}} = \frac{3\Delta f^2 E_{\text{c max}}^2}{16 R_{\text{out}} kT_{\text{s}} W^3}. \tag{7.5.47}$$

Now the *figure of merit* of FM system is

$$\text{FOM}_{\text{FM}} = \frac{(S/N)_{\text{o}}}{(S/N)_{\text{REF}}} = 1.5\left(\frac{\Delta f}{W}\right)^2 \frac{R_{\text{s}}}{R_{\text{out}}} = 1.5\, m_{\text{f}}^2 \frac{R_{\text{s}}}{R_{\text{out}}}, \tag{7.5.48}$$

where m_f is the modulation index calculated for the highest baseband frequency W. Note that this is different from the normal FM modulation index.

7.5.1 Carrier-to-noise ratio

The information on FM SNR is often expressed in another way using the carrier-to-noise ratio (CNR) as the input parameter at the detector. The CNR is similar to the $(S/N)_{\text{REF}}$ except that the total IF bandwidth is used in the place of W. So, we have

$$\left(\frac{C}{N}\right) = \frac{E_{\text{c max}}^2}{8R_{\text{s}}kT_{\text{s}}B_{\text{IF}}} \tag{7.5.49}$$

Applying Carson's rule, $B_{\text{IF}} = 2(m_{\text{f}} + 1)W$, gives

$$\left(\frac{C}{N}\right) = \frac{E_{\text{c max}}^2}{16R_{\text{s}}kT_{\text{s}}(m_{\text{f}} + 1)W}. \tag{7.5.50}$$

7.5.2 Processing gain

The ratio of the output SNR to CNR is known as the *processing gain* and is given by

$$PG = \frac{(S/N)_{\text{o}}}{(C/N)} = 3m_{\text{f}}^2(m_{\text{f}} + 1)\frac{R_{\text{s}}}{R_{\text{out}}}. \tag{7.5.51}$$

It should be noted that this result is valid only when the carrier strength is much larger than the noise and the deviation is for a sinusoidally modulated wave. When the carrier level is below the so-called *threshold level*, it is found that the output SNR worsens very rapidly.

7.6 Threshold Effect in AM Systems

The *threshold* is a value of the carrier-to-noise ratio (CNR) below which the noise performance of a demodulator deteriorates much more rapidly than the carrier-to-noise ratio. Every detector exhibits a threshold effect; below the threshold the restored message signal becomes practically useless. The physical explanation is as follows. If the carrier-to-noise ratio is high enough, then the signal dominates and the noise causes only small unwanted AM and PM. However, if the carrier-to noise ratio is small, then noise dominates which results in a complete loss of information. As a result the demodulator output does not contain the message signal at all.

7.7 Threshold Effect in FM Systems

The figure of merit of FM discussed earlier is valid only if the carrier-to-noise ratio (CNR) is high compared to unity. It has been found experimentally that as CNR decreases and falls below a threshold, each FM demodulator, either coherent or noncoherent, breaks down. At first isolated clicks are heard, and if the CNR decreases further, the clicks rapidly merge into a crackling sound. A qualitative explanation for the same is as follows. If CNR is small, then noise becomes dominant and the phasor representation and the decomposition of noise into PM and AM are not valid any more. The phase of noise is a random variable and may take any value. Recall that the FM demodulator is sensitive to the derivate of phase. When the phase of demodulator input varies suddenly by 2π due to noise, then an impulse, i.e. a click appears at the receiver output.

7.7.1 FM threshold reduction

The threshold effect exists in each FM demodulator, either coherent or noncoherent. In certain applications such as space communication, the received signal level is very low and consequently the threshold has to be reduced in practice. This can be done by a PLL FM demodulator or an FM demodulator with negative feedback (FMFB demodulator). Such circuits are referred as to *extended threshold demodulators*.

7.7.2 Comparison of noise performance in continuous wave systems

The threshold problem is more serious in FM modulation than in AM systems. The higher the modulation index, m_f, the better the FM noise performance. But the price to be paid is the wider transmission bandwidth. The high fidelity (Hi-Fi) requirements may not be satisfied by AM in the presence of noise. The noise performance of various AM and FM systems are illustrated in Figure 7.3.

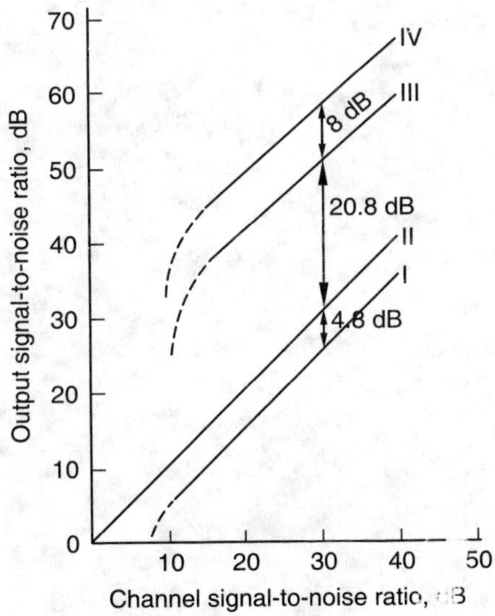

Figure 7.3 *Noise performance of AM and FM. Curve I: DSB with envelope detector,* $m_a = 1$; *curve II: DSB–SC and SSB with coherent detector; curve III: FM with* $m_f = 2$, *13 dB pre- and de-emphasis improvement included; curve IV: FM with* $m_f = 5$, *13 dB pre- and de-emphasis improvement included.*

Exercises

1. Given that the received signal power is $P_m = 0.1$ pW for a standard sinusoidal AM wave and that the baseband bandwidth is $W = 4$ kHz, calculate the output SNR in decibels. The system noise temperature is 500 K. Calculate the reference SNR for a modulation index of 0.8. Boltzmann's constant $k = 1.38 \times 10^{-23}$ J/K. Assume $R_s = R_{out}$. (Answer: 26.42 dB, 32.58 dB)
2. Calculate the figure of merit in the aforementioned case. (Answer: 24.24%)
3. A DSB-SC receiving system has an equivalent noise temperature (including antenna noise) of 1000 K and a baseband bandwidth of 4kHz. The received signal from a sinusoidally modulated DSB-SC wave is 1μ V rms across 50 Ω. Calculate (a) the output SNR, (b) the reference SNR and (c) the figure of merit. Assume $R_s = R_{out}$. (Answer: (a) 16.56 dB, (b) 16.56 dB, (c) 100%)
4. In a broadcasting communication system the transmitter power is 40 kW, the channel attenuation is 80 dB, and the noise power-spectral density is 10^{-10} W/Hz. The message signal has a bandwidth of 10^4 Hz. Find the pre-detection SNR. (Answer: 26.02 dB)

Chapter 8

Electrical Noise in Electronic Systems

In Chapter 6 we examined the mathematical theory of random variables and random processes. This will be useful in dealing with the modelling of the electrical noise in electronic systems. In layman's terms, noise is *any unwanted, uncontrollable signal*. Noise hampers the reliable performance of electronic systems by deteriorating the quality of the received signal at the information sink. Noise manifests itself in various forms: *hissing sound* in radio reception, *grass* in TV receivers and radars and so on. In several cases, noise will be found to have one of the probability density functions that we studied in chapter 6. Noise sources can be *external* or *internal*. Examples of external noises are atmospheric noise, galactic noise and man-made noise. Examples of internal noises are *shot noise* and *thermal noise*. Let us study them in greater detail.

8.1 Shot Noise

Shot noise as seen in electronic devices such as thermionic valves, semiconductor diodes and transistors is due to the discrete, random nature of current flow in these devices. Shot noise is a random fluctuation that accompanies any direct current crossing a potential barrier.

The number of electrons $\mathcal{N}(t)$ emitted in a device in the interval $[0, t]$ constitutes a discrete stochastic process, the value of which increases by 1 each time an electron is emitted. The total number of electrons emitted in the interval $[t, t+t_0]$ is denoted by x, and it can be shown to be equal to

$$x = \mathcal{N}(t+t_0) - \mathcal{N}(t).$$

The mean value of electrons emitted between times t and $t+t_0$ is given by

$$E[x] = \lambda t_0, \tag{8.1.1}$$

where λ is a constant called the *rate of the process*. It is known that x follows a Poisson distribution, and the probability that k electrons are emitted in the interval $[t, t+t_0]$ is given by

$$P\{x = k\} = \frac{(\lambda t_0)^k}{k!} e^{-\lambda t_0}, \quad k = 0, 1, 2, \cdots. \tag{8.1.2}$$

The mean of the total current flowing through the device $X(t)$, which is a random process, is given by

$$\mu_X = \lambda \int_{-\infty}^{\infty} h(t)dt \tag{8.1.3}$$

where λ is the rate of the process and $h(t)$ the waveform of the current pulse. The autocovariance function of $X(t)$ is given by

$$C_X(\tau) = \lambda \int_{-\infty}^{\infty} h(t)h(t+\tau)dt, \tag{8.1.4}$$

This result is known as *Campbell's theorem.* When the waveform $h(t)$ consists of rectangular pulses of amplitude A and duration T, we have

$$\begin{aligned} \mu_X &= \lambda AT \\ C_X(\tau) &= \begin{cases} \lambda A^2(T-|\tau|), & |\tau| < T \\ 0, & |\tau| \geq T. \end{cases} \end{aligned} \tag{8.1.5}$$

The mean square noise component of shot noise is proportional to the dc current, and for most devices the mean-square shot noise current is given by

$$I_{\mathrm{n}}^2 = 2I_{\mathrm{dc}}q_{\mathrm{e}}B_{\mathrm{n}} \ \text{ampere}^2, \tag{8.1.6}$$

where I_{dc} is the direct current in amperes, q_{e} the magnitude of the electron charge ($= 1.6 \times 10^{-19}$ C), and B_{n} the equivalent noise bandwidth in hertz.

Example 8.1 Calculate the shot noise component of current present in a direct current of 100 mA flowing across a semiconductor junction given that the effective noise bandwidth is 1 MHz.

Solution

$$\begin{aligned} I_{\mathrm{n}}^2 &= 2 \times 10^{-1} \times 1.6 \times 10^{-19} \times 10^6 \\ &= 3.2 \times 10^{-14} \ \text{ampere}^2. \\ I_{\mathrm{n}} &= 0.1789 \times 10^{-6} \ \text{amperes}. \end{aligned}$$

8.2 Thermal Noise

Thermal noise is the electrical noise arising from the random motion of electrons in a conductor. The mean square value of the thermal noise ${E_{\mathrm{n}}}^2$ is given by

$$E_{\mathrm{n}}^2 = 4kTR\Delta f \ \text{volts}^2, \tag{8.2.7}$$

where k is the Boltzmann's constant ($= 1.38 \times 10^{-23}$ J/K), T is the absolute temperature in degree Kelvin, and R is the resistance in Ωs. Thus, we can model a noisy resistor by the *Thevenin equivalent circuit* consisting of a noise voltage generator of mean square value E_{n}^2 in series with a noiseless resistor. Alternatively, we can use the *Norton equivalent circuit* consisting of a noise current generator in parallel with a noiseless conductance. The mean square value of the noise current source is given by

$$I_{\mathrm{n}}^2 = 4kTG\Delta f \quad \text{ampere}^2, \tag{8.2.8}$$

where $G = \frac{1}{R}$.

8.2.1 Equivalent noise bandwidth

When a resistor R is connected to the input of an LC filter, the total mean square output voltage is given by

$$V_{\mathrm{n}}^2 = \int_0^{\infty} 4RkT|H(f)|^2 \, df \tag{8.2.9}$$

$$= 4RkT \times (\text{area under the } |H(f)|^2 \text{ curve}). \tag{8.2.10}$$

Now the total mean square voltage at the output can be stated as $V_{\mathrm{n}}^2 = 4RkTB_{\mathrm{n}}$, and equating with (8.2.10), we have the expression for *equivalent noise bandwidth, B_{n}*:

$$B_{\mathrm{n}} = \int_0^{\infty} |H(f)|^2 \, df$$

$$= \text{area under the } |H(f)|^2 \text{ curve}. \tag{8.2.11}$$

8.3 White Noise

The noise analysis in communications systems is based on an idealised form of noise known as *white noise*. The power spectral density (PSD) of white noise is independent of frequency, and hence the name.

8.3.1 Equivalent noise temperature

We can express the PSD of white noise as

$$S_{\mathrm{W}}(f) = \frac{N_{\mathrm{O}}}{2}, \quad \text{for} \quad -\infty \leq f \leq \infty. \tag{8.3.12}$$

The dimensions of N_{O} are in watts per hertz. N_{O} may be expressed as

$$N_{\mathrm{O}} = kT_e, \tag{8.3.13}$$

where k is the Boltzmann's constant and T_e is the *equivalent noise temperature* of the receiver. The definition of equivalent noise temperature is as follows.

The equivalent noise temperature is the temperature at which a noisy resistor has to be maintained such that by connecting the resistor to the input of a noiseless version of the system, it produces the same available noise power at the output of the system as that produced by all the sources of noise in the actual system.

As the autocorrelation function is the inverse Fourier transform of the power spectral density, we have

$$R_W(\tau) = \frac{N_0}{2}\delta(\tau). \tag{8.3.14}$$

8.4 Partition Noise

Partition noise occurs whenever current has to divide between two or more electrodes and results from the random fluctuations in the division. It would be expected that a diode is less noisy than a transistor (other factors being equal) if the third electrode draws current. The spectrum for partition noise is flat.

8.5 Low Frequency or Flicker Noise

Flicker noise becomes apparent below frequencies of the order of a few kilohertz. It is also known as $\frac{1}{f}$ noise. In vacuum tubes it arises from the slow changes in the oxide structure of the oxide-coated cathodes and from migration of impurity ions through the oxide. In semiconductors, flicker noise arises from fluctuations in the carrier densities (holes and electrons), which in turn give rise to fluctuations in the conductivity of the material.

8.6 Burst Noise

This is another low-frequency noise observed in transistors. The noise appears as a series of bursts of two or more levels. When present in an audio system, the noise produces *popping sounds*, and for this reason it is also known as *popcorn noise*. The spectral density of burst noise is known to increase as the frequency decreases.

8.7 Signal-to-Noise Ratio

In a communications system the *signal-to-noise ratio* (SNR) is more important than the absolute value of noise. We can define it as

$$\frac{S}{N} = \frac{P_{\text{sig}}}{P_{\text{noi}}} = \frac{V_{\text{sig}}^2}{V_{\text{noi}}^2}. \tag{8.7.15}$$

SNR is expressed in decibels as $\log_{10}\left(\frac{S}{N}\right) = 20\log_{10}\left(\frac{V_{\text{sig}}}{V_{\text{noi}}}\right)$.

8.8 Representation of Narrowband Noise

The receiver of a communications system usually includes some provision for *pre-processing* of the received signal. This may take the form of a narrowband filter whose response is just large enough to pass the modulated component of the received signal but not so large as to admit excessive noise through the receiver. The noise process operating at the output of such a filter is called *narrowband noise*. Depending on the application of interest, we have two representations for narrowband noise, which are discussed in the following subsections.

8.8.1 In-phase and quadrature components

Consider a narrowband noise $n(t)$ of bandwidth $2B$ centred at frequency f_c. We can represent $n(t)$ as

$$\begin{aligned} n(t) &= A(t)\,\cos\left[2\pi\, f_c t + \psi(t)\right] \\ &= A(t)\,\cos\left[\psi(t)\right]\,\cos\left[2\pi\, f_c t\right] - A(t)\,\sin\left[\psi(t)\right]\,\sin\left[2\pi\, f_c t\right] \\ &= n_I(t)\,\cos\left[2\pi\, f_c t\right] - n_Q(t)\,\sin\left[2\pi\, f_c t\right], \end{aligned} \tag{8.8.16}$$

where $n_I(t)$ is the *in-phase component* and $n_Q(t)$ the *quadrature component*.

Given the narrowband noise $n(t)$, we can extract the in-phase and quadrature components as shown in Figure 8.1. Also, we can reconstruct a narrowband noise process from its components as shown in Figure 8.1.

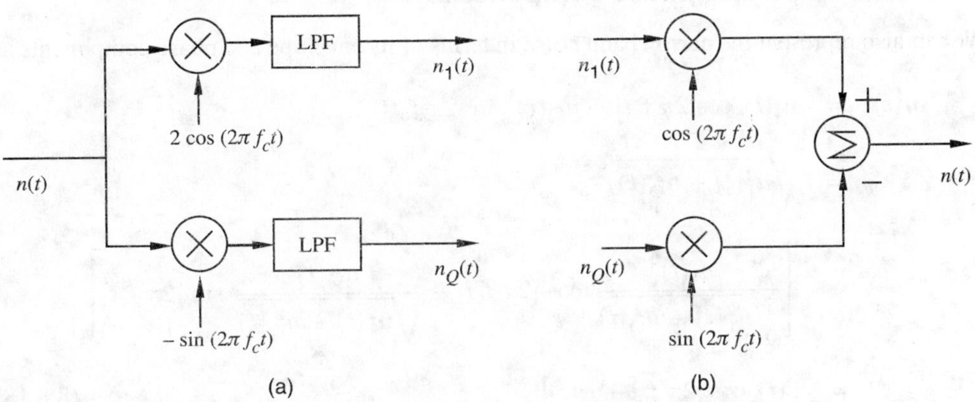

Figure 8.1 *Extraction of components from* $n(t)$

The important properties of in–phase and quadrature components are listed as follows.

1. The in-phase and quadrature components have zero mean.
2. If the narrowband noise $n(t)$ is Gaussian, then its in-phase and quadrature components are jointly Gaussian.
3. If the narrowband noise is stationary, then its in-phase and quadrature components are jointly stationary.

4. Both the components have the same power spectral density, which is related to the PSD $S_N(f)$ of the narrowband noise $n(t)$ as

$$S_{N_I}(f) = S_{N_Q}(f) = \begin{cases} S_N(f-f_c) + S_N(f+f_c), & -B \le f \le B \\ 0, & \text{otherwise,} \end{cases}$$

where its is assumed that $S_N(f)$ occupies the frequency interval $f_c - B \le |f| \le f_c + B$ and $f_c > B$.

5. The in-phase and quadrature components have the same variance as the narrowband noise.

6. The cross-spectral density of the in-phase and quadrature components of narrowband noise is purely imaginary, as

$$\begin{aligned} S_{N_I N_Q}(f) &= -S_{N_Q N_I}(f) \\ &= \begin{cases} j[S_N(f+f_c) - S_N(f-f_c)], & -B \le f \le B \\ 0, & \text{otherwise.} \end{cases} \end{aligned} \tag{8.8.17}$$

7. If the narrowband noise is Gaussian and its PSD $S_N(f)$ is symmetric about the mid-band frequency f_c, then the in-phase and quadrature components are statistically independent.

8.8.2 Envelope and phase components

We can also represent the narrowband noise in terms of its envelope and phase components as

$$\begin{aligned} n(t) &= n_I(t)\cos[2\pi f_c t] - n_Q(t)\sin[2\pi f_c t] \\ &= \sqrt{n_I^2(t) + n_Q^2(t)} \\ &\quad \left[\frac{n_I(t)}{\sqrt{n_I^2(t) + n_Q^2(t)}} \cos[2\pi f_c t] - \frac{n_Q(t)}{\sqrt{n_I^2(t) + n_Q^2(t)}} \sin[2\pi f_c t] \right] \\ &= r(t)\cos[2\pi f_c t + \psi(t)], \end{aligned} \tag{8.8.18}$$

where

$$r(t) = \sqrt{[n_I^2(t) + n_Q^2(t)]} \quad \text{and} \tag{8.8.19}$$

$$\psi(t) = \tan^{-1}\left[\frac{n_Q(t)}{n_I(t)}\right]. \tag{8.8.20}$$

Note that N_I and N_Q are independent Gaussian random variables of zero mean and variance σ^2, and we may write their joint probability function distribution (PDF) as

$$f_{N_I N_Q}(n_I, n_Q) = \frac{1}{2\pi\sigma^2} \exp\left(-\frac{n_I^2 + n_Q^2}{2\sigma^2}\right). \tag{8.8.21}$$

Defining the transformations

$$n_I = r \cos \psi(t). \tag{8.8.22}$$

$$n_Q = r \sin \psi(t), \tag{8.8.23}$$

we can express the joint PDF of R and ψ as

$$f_{R,\psi}(r, \psi) = \frac{r}{2\pi\sigma^2} \exp\left(-\frac{r^2}{2\sigma^2}\right). \tag{8.8.24}$$

The random variables R and ψ are statistically independent. Hence, we can write

$$f_\psi(\psi) = \begin{cases} \frac{1}{2\pi}, & 0 \le \psi \le 2\pi \\ 0, & \text{elsewhere,.} \end{cases} \tag{8.8.25}$$

and

$$f_R(r) = \begin{cases} \frac{r}{\sigma^2} \exp\left(-\frac{r^2}{2\sigma^2}\right), & r \ge 0 \\ 0, & \text{elsewhere,.} \end{cases} \tag{8.8.26}$$

where σ^2 is the variance of the original narrowband noise $n(t)$.

8.9 Sine Wave Plus Narrowband Noise

Let us assume that we add a sinusoidal wave $A\cos(2\pi f_c t)$ to the narrowband noise $n(t)$, where A and f_c are both constants. Assume that the frequency of the sinusoidal wave is the same as the nominal carrier frequency of the noise. Then we can write a sample function as

$$\begin{aligned} x(t) &= A \cos(2\pi f_c t) + n(t) \\ &= A \cos(2\pi f_c t) + n_I(t) \cos(2\pi f_c t) - n_Q(t) \sin(2\pi f_c t) \\ &= n_I'(t) \cos(2\pi f_c t) - n_Q(t) \sin(2\pi f_c t) \\ &= \sqrt{[n_I'(t)]^2 + n_Q^2(t)} \\ &\quad \left[\frac{n_I'(t)}{\sqrt{[n_I'(t)]^2 + n_Q^2(t)}} \cos(2\pi f_c t) - \frac{n_Q(t)}{\sqrt{[n_I'(t)]^2 + n_Q^2(t)}} \sin(2\pi f_c t)\right] \\ &= r(t) \cos(2\pi f_c t + \psi(t)), \end{aligned} \tag{8.9.27}$$

where

$$n_I'(t) = A + n_I(t), \tag{8.9.28}$$

$$r(t) = \sqrt{[n_I'(t)]^2 + n_Q^2(t)}, \tag{8.9.29}$$

$$\psi(t) = \tan^{-1}\left[\frac{n_Q(t)}{n_I'(t)}\right]. \tag{8.9.30}$$

Now $r(t)$ is the envelope of $x(t)$ and $\psi(t)$ is its phase. Assuming $n(t)$ to be Gaussian with zero mean and variance σ^2, we can state:

1. Both $n_I'(t)$ and $n_Q(t)$ are Gaussian and statistically independent.
2. The mean of $n_I'(t)$ is A and that of $n_Q(t)$ is zero.
3. The variance of both $n_I'(t)$ and $n_Q(t)$ is σ^2.

Now the joint PDF of random variables R and ψ is given by

$$f_{R,\psi}(r,\psi) = \frac{r}{2\pi\sigma^2}\exp\left(-\frac{r^2 + A^2 - 2Ar\cos\psi}{2\sigma^2}\right). \tag{8.9.31}$$

To determine the PDF of random variable R, we integrate (8.9.31) with respect to ψ. Therefore, we have

$$\begin{aligned} f_R(r) &= \int_{\psi=0}^{2\pi} f_{R,\psi}(r,\psi)d\psi \\ &= \frac{r}{2\pi\sigma^2}\exp\left(-\frac{r^2 + A^2}{2\sigma^2}\right)\int_0^{2\pi}\exp\left(\frac{Ar}{\sigma^2}\cos\psi\right)d\psi. \end{aligned} \tag{8.9.32}$$

Note that $\frac{1}{2\pi}\int_0^{2\pi}\exp(x\cos\psi)d\psi = I_0(x)$, the *modified Bessel function of the first kind of zero order*. Then by letting $x = Ar/\sigma^2$, we can write (8.9.32) as

$$f_R(r) = \frac{r}{\sigma^2}\exp\left(-\frac{r^2 + A^2}{2\sigma^2}\right) I_0\left(\frac{Ar}{\sigma^2}\right). \tag{8.9.33}$$

This PDF is called the *Rician distribution.* A plot of the Rician distribution is given in Figure 8.2. The MATLAB script to generate the plot is given below.

```
%% MATLAB script m-file to generate and
% plot the Rician p.d.f. for various values of
%  a and v...
clf;
clear all;
 v=[0:.01:10];
for a=0:5
```

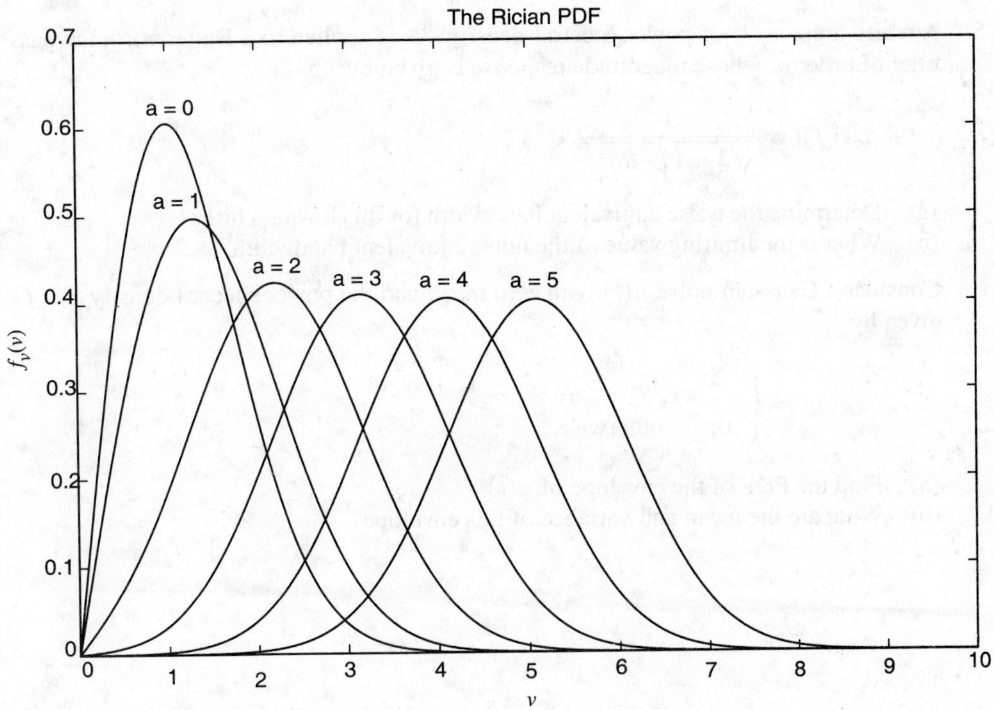

Figure 8.2 *The Rician probability distribution*

```
rpdf=  v.*exp(-(v.^2+a^2)/2).*besseli(0,a*v);
plot(v,rpdf);   hold on;
end;
xlabel('\nu'); ylabel('f_{\nu}(\nu)');
title('The  Rician p.d.f.');
hold off;
```

Exercises

1. The thermal noise from a resistor is $4 \times 10^{-17} W$ for a given bandwidth and at a temperature of $20°C$. What will be the noise power at 70 K? (Answer: $\frac{4\times10^{-17}\times70}{293} = 9.556 \times 10^{-18}$ W)
2. Given two resistors $R_1 = 10$ k Ω and $R_2 = 15$ k Ω, calculate the thermal noise voltage generated at a temperature of 290 K and bandwidth of 20 MHz when (a) R_1 in series with R_2, (b) R_1 in parallel with R_2. (Answer: (a) 89.465 μV (b) 43.829 μV)
3. The noise generated by a 2000 Ω resistor can be represented by a 4 nA current source in parallel with a 2000 Ω. Find out the equivalent voltage source representation. (Answer: 8 μV in series with 2000 Ω)
4. Calculate the shot noise current present in a direct current of 13 mA for a noise bandwidth of 7 MHz. (Answer: 0.171 μA)

5. A white noise $w(t)$ of power spectral density $\frac{N_o}{2}$ is applied to a Butterworth lowpass filter of order n, whose magnitude response is given by

$$|H(f)| = \frac{1}{\sqrt{1 + (f/f_o)^{2n}}}.$$

   (a) Determine the noise equivalent bandwidth for this lowpass filter.
   (b) What is the limiting value of the noise equivalent bandwidth?

6. Consider a Gaussian noise $n(t)$ with zero mean and the power spectral density $S_N(f)$ given by

$$S_N(f) = \begin{cases} \frac{N_o}{2}, & f_c - B \leq |f| \leq f_c + B \\ 0, & \text{otherwise.} \end{cases}$$

   (a) Find the PDF of the envelope of $n(t)$.
   (b) What are the mean and variance of this envelope?

Appendix

This section contains a few sets of sample questions on analong communications taken from the B.Tech degree examinations conducted by Kerala University.

MODEL QUESTION PAPER - I (2005)

Time: Three Hours | Fourth Semester | Maximum Marks: 100

Answer all questions from Part A. Each question carries 4 marks. Answer any two questions from each module of Part B. Each question carries 10 marks.

PART – A

1. Explain the functions of the SLIC in telephone systems.
2. Obtain the frequency spectra in the case of AM, DSB-SC and SSB.
3. Explain the block diagram of a superheterodyne receiver.
4. Explain the functioning of a balanced modulator to generate DSB-SC.
5. Explain the functioning of an FM detector using PLL.
6. Explain the functioning of a ratio detector.
7. Derive the expression for SNR in an AM system.
8. A coin is tossed three times and the random variable X denotes the total number of heads that appear. The probability of a head in one toss is p. (a) What values can the random variable X take? (b) What is its probability mass function?
9. Obtain the mean and variance of a Poisson random variable.
10. State any four properties of power spectral density. $(10 \times 4 = 40)$

PART – B

MODULE – I $(2 \times 10 = 20)$

11. With the help of a block diagram, explain the functioning of two-wire and four-wire circuits. Explain the working of a digital exchange.
12. Differentiate between low level and high level modulators. What are the merits of double conversion receivers?
13. What is the principle of SSB modulation? State its advantages, with typical numerical examples.

MODULE – II $(2 \times 10 = 20)$

14. Explain the need for pre-emphasis and de-emphasis in FM systems.
15. Explain how PM and FM are related. Show how one can be generated from the other.
16. What are the different noise sources in FM receivers? Compare the noise performance of AM and FM.

MODULE – III $(2 \times 10 = 20)$

17. Let X and Y be independent random variables with $f_X(x) = \alpha \exp(-\alpha x), \; x > 0$ and 0 otherwise; and $f_Y(y) = \beta \exp(-\beta y), \; y > 0$ and 0 otherwise, where α and β are assumed to be positive constants. Find the PDF of $X + Y$.
18. Derive the figure of merit as regards to SNR in DSB–SC AM system.
19. Show that the output of an LTI system when the input is a normal random variable, is also a normal random variable.

QUESTION PAPER - II (2005)

Time: Three Hours | Fourth Semester | Maximum Marks: 100

Answer all questions from Part A. Each question carries 4 marks.
Answer any two questions from each module of Part B. Each question carries 10 marks.

PART – A

1. Explain the blocks (components) of a telephone set.
2. Define modulation index and its importance.
3. Discuss any one method of AM demodulation.
4. Explain how angle modulation is generated.
5. Explain the frequency vs. gain characteristics of pre-emphasis filter and de-emphasis filter.
6. Explain the advantages of FM over AM.
7. List the properties of cross correlation.
8. Find the power spectral density of sinusoid, $g(t) = A\cos(\omega_0 t + \theta)$.
9. Explain S/N in analog communication system.
10. Discuss the representation of narrowband noise in terms of envelope and phase components. $(10 \times 4 = 40)$

PART – B

MODULE – I $(2 \times 10 = 20)$

11. Explain with help of a neat sketch the working of superheterodyne receiver. Also, how tracking is carried out.
12. What is SSB suppressed carrier modulation? What are its advantages and disadvantages with respect to AM? Explain through diagrams.
13. Discuss the circuit of AGC and its practical significance.

MODULE – II (2 × 10 = 20)

14. (a) Describe the generalised concept of angle modulation.
15. (b) An angle modulated signal with carrier frequency $\omega_c = 2\pi \times 10^5$rad/sec is described by the equation:

$$\phi_m(t) = 10 \cos\,(\omega_c t + 5 \sin 3000t + 10 \sin 2000t)$$

(a) Find the frequency deviation.
(b) Find the phase deviation.

16. Explain a method of FM detection and also how frequency is stabilised.
17. Write short notes on

(a) FM broadcast.
(b) Noise in AM and FM detection.

MODULE – III (2 × 10 = 20)

18. (a) Define mean, correlation and covariance.
(b) Show that the random process $X(t) = A\cos(\omega_c t + \theta)$, where θ is a random variable distributed in the range $(0, 2\pi)$, is a wide-sense stationary process.
19. (a) Show that PSD is the Laplace transform of the correlation function.
(b) Explain ergodic process.
20. Write short notes on:

(a) Central-limit theorem and its properties
(b) Noise in communication system

QUESTION PAPER - III (2006)

Time: Three Hours Fourth Semester Maximum: 100 Marks

Answer all questions from Part A. Each question carries 4 marks.
Answer any two questions from each module of Part B. Each question carries 10 marks.

PART – A

1. Briefly explain DSB-SC modulation with diagrams.
2. What is sensitivity in the case of receivers?
3. Draw the block diagram of a SSB demodulator. Plot the frequency domain representation of output waveforms of coherent detection of SSB (single side band).
4. What are the different forms of angle modulation schemes? Briefly explain each.
5. Compare the different modulation schemes of FM.
6. What is *capture effect* in FM?
7. Write the axioms of probability theory.
8. Explain how the ergodicity of a random process can be tested.
9. Show that $R_\nu(\tau) = R_\nu(-\tau)$, where $R_\nu(\tau)$ is autocorrelation function.

10. Compare autocorrelation with cross correlation. [10 × 4 = 40]

PART – B

MODULE -I (2 × 10 = 20)

11. With the help of diagrams, explain how AM signals can be detected.
12. With the help of diagrams, explain coherent detection of DSBSC waves.
13. Explain with diagrams the frequency discrimination method of generation of SSB wave.

MODULE -II (2 × 10 = 20)

14. Derive a relation for figure of merit of an AM receiver.
15. Explain, with the help of diagrams, pre-emphasis and de-emphasis in FM.
16. Derive a relation for the spectrum of a wideband frequency modulated wave. What are its properties?

MODULE -III (2 × 10 = 20)

17. Find the frequency function and cumulative distribution function for transmitting 3 digit messages over a noisy channel. The channel error probability $P(E) = \frac{2}{5}$ per digit. The errors are statistically independent from digit to digit. Also plot the frequency functions and CDFs.
18. What are different probability models of a continuous random variable (rv). Derive an expression for PDF using any one of the probability models of a continuous rv.
19. Show that output signal $y(t)$ of a filtered random signal is wide sense stationary (assume any input properly).

QUESTION PAPER - IV (2004)
UNIVERSITY OF KANNUR

Time: Three Hours | Fifth Semester | Maximum: 100 Marks

PART – A

*Answer **all** questions.*

1. (a) Define mean and correlation functions.
   (b) The PDF of a random variable X is given as $f_x(X)$. A random variable Y is defined as $Y = aX + b$, where $a < 0$. Determine the PDF of Y in terms of the PDF of X.
   (c) Define noise figure and noise equivalent temperature.
   (d) Write short note on narrow bandpass noise.
   (e) Compare the following modulation schemes: AM, DSBSC, SSB.
   (f) Derive an expression for total power in an AM system.
   (g) Explain deviation ratio and modulation index in FM.
   (h) A sinusoidal modulating wave of amplitude 5 V and frequency 1 kHz is applied to a frequency modulator. The frequency sensitivity of the modulator is 50 Hz/V and carrier frequency is 100 kHz. Calculate the frequency deviation and bandwidth.

(8 × 5 = 40 marks)

PART – B

Answer either (a) or (b).

2. (a) (i) The PDF of a continuous random variable is said to have Laplace distribution given by

$$f_X(X) = \frac{1}{2}e^{-|x|}; \quad -\infty < x < \infty.$$

Find the mean and the variance of the rv. (10 marks)

(ii) Determine the probability that the Gaussian random variable X lies in the interval

$$m_x - K\sigma_x < X\, m_x + K\sigma_x,$$

where $x = 3$. (5 marks)

or

(b) (i) State and prove Wiener–Khinchine–Einstein theorem (8 marks)

(ii) Define Gaussian random process. Explain its properties. (7 marks)

3. (a) (i) Derive the expression for the overall noise figure and the overall noise equivalent temperature for noise networks connected in cascade. (10 marks)

(ii) A receiver with a noise figure of 10 dB is fed by a LNA that has a gain of 60 dB and a noise temperature of 80 K. Calculate the noise temperature of the receiver and the overall noise temperature. Assume a room temperature of 25° C. (5 marks)

or

(b) (i) Three telephone links, each having an SNR of 44 dB, are connected in tandem. What is the overall SNR? A fourth link is now added, which has SNR of 34 dB. What is the overall SNR? (9 marks)

(ii) Three 5 K resistors are connected in series at room temperature ($KT = 4 \times 10^{-21}$) and an effective noise bandwidth of 1 MHz. Determine (1) the noise voltage appearing across each resistor; (2) the noise voltage appearing across series combination; and (3) the noise voltage appearing across parallel combination. (6 marks)

4. (a) (i) Illustrate time and frequency domain characteristics of standard amplitude modulation produced by a single tone frequency component. Also, obtain the equation for total power, upper-side frequency power, lower-side frequency power. (8 marks)

(ii) Explain the generation of AM using ring modulator for sinusoidal modulating wave with necessary equations. (7 marks)

or

(b) (i) Explain the generation of SSB using phase discrimination method. (5 marks)

(ii) Explain the generation of DSBSC using balanced modulator and demodulation of DSBSC using coherent detector. (10 marks)

5. (a) (i) Explain direct and indirect method of FM generation. (10 marks)
   (ii) The equation for an FM wave is given by

$$s(t) = 10 \sin (5.7 \times 10^8 \, t + 5 \sin 12 \times 10^3 \, t).$$

Calculate the carrier frequency, modulating frequency, modulation index, frequency deviation and power dissipated in a 100 Ω resistor. (5 marks)

or

(b) (i) Explain the operation of a balanced slope detector with its relevant circuit diagram and frequency response curve. (8 marks)
   (ii) Write short notes on: (a) pre-emphasis. (b)de-emphasis. (7 marks)

Bibliography

[1] Simon Haykin, *Communication Systems*, 4th edition, John Wiley & Sons, 2001.

[2] George Kennedy and Bernard Davis, *Electronic Communication Systems*, 4th edition, Tata McGraw-Hill Publishing Company Limited, 1999.

[3] B.P. Lathi, *Modern Digital and Analog Communication Systems*, 3rd edition, Oxford University Press, 2004.

[4] John G.Proakis and Masoud Salehi, *Communication Systems Engineering*, 2nd edition, Pearson Education, 2002.

[5] Robert J. Schoenbeck, *Electronic Communications Modulation and Transmission*, Universal Book Stall, 1992.

[6] A. Papoulis and S. Unnikrishna Pillai, *Probability, Random Variables and Stochastic Processes*, 4th edition, Tata McGraw-Hill, 2003.

[7] Martin S. Roden, *Analog and Digital Communication Systems*, 5th edition, Shroff Publishers, 2002.

[8] Dennis Roddy and John Coolen, *Electronic Communications*, 4th edition, Prentice Hall of India, 2000.

[9] Michel C. Jeruchim, Philip Balaban, and K. Sam Shanmugan, *Simulation of Communication Systems*, Plenum Press, 1992.

[10] William J. Palm-III, *Introduction to MATLAB 7 For Engineers*, McGraw-Hill Inc., 2004.

[11] John G.Proakis and Masoud Salehi, *Contemporary Communication Systems Using MATLAB*, Vikas Publishing House, 2002.

[12] A. Bruce Carlson, Paul B. Crilly, and Janet C. Rutledge, *Communication Systems*, 4th edition, McGraw-Hill Inc., 2002.

[13] Peyton Z. Peebles, Jr., *Probability, Random Variables and Random Signal Principles*, 4th edition, Tata McGraw-Hill, 2003.

[14] Erwin Kreyszig, *Advanced Engineering Mathematics*, 8th edition, Wiley Eastern, 2006.

Index